FIRE
AND
ICE

THE ORIGIN
OF LIFE ON EARTH

PETER ROY CLEMENTS

Hardcover: 978-1-963050-70-7
Paperback: 978-1-963050-63-9
eBook: 978-1-963050-64-6
Library of Congress Control Number: 2023923192

Ordering Information:

Prime Seven Media
518 Landmann St.
Tomah City, WI 54660

Printed in the United States of America

Figure 1 (Cover illustration), Hot spring amid snow covered ground.

TABLE OF CONTENTS

Origin of life in fire and ice.

By Peter R Clements, PhD.
University of Adelaide,
Adelaide, South Australia
Australia
Email: peter.clements@adelaide.edu.au

It is often said that all the conditions for the first production of a living organism are now present, which could ever have been present.— But if (& oh what a big if) we could conceive in some warm little pond with all sorts of ammonia & phosphoric salts,—light, heat, electricity &c present, that a protein compound was chemically formed, ready to undergo still more complex changes, at the present day such matter w^d be instantly devoured, or absorbed, which would not have been the case before living creatures were formed.

Charles Darwin Letter to J. D. Hooker, 1 Feb [1871]

ABSTRACT

A simple and direct path is proposed towards the formation of the first organism, LUCA (the last universal common ancestor), which is described using observations found in the scientific literature. The path is proposed to take place in the environment provided by hot springs and in the presence of snow and ice thus creating the "FIRE and ICE" conditions. The path proposed is one which is plausible given that the experiment cannot be re-run. The path consists of the following steps: delivery of water and organic compounds to the early Earth via comet and asteroid impacts, formation of vesicles in geysers, entrapment of clay particles, amino acids and other ingredients in the vesicles, formation of template-directed peptides, elongation to peptides with catalytic activity, association of catalytic peptides with aromatic compounds including purines and pyrimidines, peptide catalyzed development of nucleotides, polymerization of nucleotides to RNA, the RNA world, stereochemical association with amino acids and peptides, the RNA-peptide world, allow the RNA world to develop a 'code' which develops into the genetic code and the advent of LUCA. This is not an extensive literature review but presents as simplified a path towards the first organism as can be envisaged using all steps found in the literature.

PROLOGUE

Solar system neighbours

Life on Earth consists of millions of species including all forms from bacteria, fungi through to plants and animals. There are forms of life populating every possible environment on Earth. From the ocean depths, where many strange creatures live, to the Earth's atmosphere where there are microorganisms floating about. There are algae that can live below zero temperatures under ice sheets and there are bacteria that can live at near boiling temperatures in hot springs [1].

So, life has been highly successful on this planet filling every possible niche and yet close by, our nearest solar system neighbour, the moon, is entirely devoid of life. What about the rest of the solar system?

While efforts are being made to find possible life on Mars, a planet close to us, there is still no direct proof of life there either. Mars is quite a bit smaller than Earth, is significantly colder and it lacks an atmosphere capable of sustaining life as we know it. Yet there could still have been non oxygen–dependent life there. From extensive data sent by the probes roaming the surface of Mars there is no life so far and this is despite clear evidence that flowing water has been present on Mars at some time in its history.

Venus, another close solar system neighbour, has an atmosphere so thick and hot that it would melt lead and this condition alone rules out any thought of life taking hold there. Even if there once was life there it would have been long obliterated by the current highly acidic and very hot atmospheric conditions there. The atmosphere is rich in carbon dioxide affording a runaway greenhouse effect and heating the atmosphere to very high temperatures.

Other planets are either too close to the sun as in Mercury, or too far away and therefore too cold as in Jupiter, Saturn, Uranus and Neptune. Some moons of these planets are thought to contain some liquid water under their icy surfaces such as Jupiter's Europa and possibly Saturn's Enceladus.

Why does Earth support life while other planets in our solar system do not? The Goldilocks principle.

Earth is home to life because it is just the right distance from the sun. By virtue of having an atmosphere containing gases including carbon dioxide, nitrogen and water vapour, enough solar heat is trapped to allow water to exist in mostly its liquid state. Earth's atmosphere also contains about 21% oxygen which generates ozone in the upper layers and this protects life on Earth from exposure to damaging levels of ultraviolet radiation. The warming provided by the other gases provides and maintains a narrow range of temperatures for life to exist. This is a level of global warming, known as the greenhouse effect, that is necessary for life, as opposed to the human-created artificially high levels now threatening life on Earth by enhancing the level of warming at too fast a rate for biological systems to adapt. Without the level of climate moderation that the atmosphere provides, Earth would be too hot during the day and too cold during the night for life to exist. At the same distance from the sun as the Earth, with no atmosphere, the moon adequately demonstrates this.

The circumstances of Earth's position in the solar system, and the contribution made by its atmosphere, which gives rise to the ideal life-supporting conditions has been called the Goldilocks principle. Earth is not too hot, nor too cold, but just right for life. It is a subject of philosophical argument whether this principle is a product of our human desire to fit everything in nature into our (anthropocentric) world view or whether there is some other, yet undiscovered, process at work which results in these ideal conditions; or is it just a set of chance happenings?

Life on Earth, running the program backwards.

What Charles Darwin and Alfred Wallace showed in their theory of Evolution, which they arrived at independently, was that all the life forms on Earth at the present time have evolved from previous life forms [2]. This is now one of the most scientifically supported theories that science has come up with and it is a guiding principle for all biological studies. Ultimately the strongest evidence supporting the theory is best found by looking at DNA encoding genes from all species. There are genes encoding for many important proteins which are nearly identical (highly conserved amino acid sequences) across all species. Indeed, the mechanism by which the genes are translated into proteins, the Genetic Code, is close to identical for all species on Earth (some very primitive species have slight variations in the code-a clue for its origin!) Therefore, knowing how identical for all species is the process by which DNA is replicated (copied) and passed on to new cells, and how identical is the process by which genes for all species are translated into proteins, there can be no other credible origin for evolution than that all species on Earth have arisen from a common ancestor.

If we run the program backwards by which those life forms arose we find that looking at DNA sequences, humans' closest relative is the chimpanzee but, to correct a common misconception, the chimpanzee is not a species from which humans arose. Humans, bonobos (another chimpanzee-like species) and chimpanzees had a common ancestor at one time from which we all diverged only as recently (in geological terms) as a few million years ago. Perhaps this divergence occurred more recently, since the oldest known identifiably human fossils found in Morocco recently are dated at 300,000 years old. The human/chimpanzee/bonobo ancestor no longer exists but it in turn had an ancestor and if we wind the tape backwards through possible precursor ape species we arrive at a lemur-like mammal that gave rise to the apes and all that family of mammals.

Richard Dawkins has explored these lineages exhaustively and illustrated them well in a book called 'The Ancestor's Tale' [3]. Going further back we encounter a species that developed fur and a warm-blooded metabolism to split from its cold blooded dinosaur–like reptilian ancestors. It was a mammalian species that survived the asteroid impact in the Yucatan peninsula which the dinosaurs did not survive. Reptile lineage can be traced back through their antecedents, the amphibians, the first creatures to have both an aquatic and a land based existence. Their descendants today are frogs and salamanders. Prior to amphibians were lungfish which were of course evolved from fish. There are still lungfish in waters today and these can be considered living fossils in that they have examples in the fossil record as well as living examples today. The most archetypical of these living fossils is the coelacanth, rediscovered off the east coast of Africa in 1938, not so long ago.

The species which came before fish were arthropods such as crustaceans, crabs and a vast number of insect species and their precursor species are creatures like trilobites of which only fossil evidence remains.

Mass extinctions: A fascinating study of the earliest arthropods can be found in looking at Precambrian fauna. Fossils of this era were first found in a deposit called the Burgess Shales at Yoho National Park in Alberta, Canada. The story of the Burgess Shale fossils and their significance were illustrated in a book called "Wonderful Life" by Stephen Jay Gould [4]. In this absorbing book some interesting observations are made about the variety of different body plans, most with exoskeletons (such as found in crabs and lobsters), that once existed and how they were largely obliterated by a mass extinction event. Probably an asteroid impact, the evidence for which is still controversial. It is likely that the dominant remaining species after that event were the trilobites. Gould points out that there is no way to predict which species will survive such a mass extinction event, but the end result is that all species that follow are derived from those remaining species. This means that all body plans are reduced to being derived from the ones remaining which are contained in those surviving species. That might be only a few percent of the variety that existed before the mass extinction event.

Mass extinction events have happened a number of times in Earth's prehistory and each time the species that are left will diverge to fill all the ecological niches left vacant by the mass extinctions.

Fossils of earliest life

Predating these fossils with exoskeletons were a group of more primitive worm-like creatures in a fossil deposit in South Australia in the Ediacaran area of the Flinders Ranges. Discovered by Reg Sprigg, these fossils from 635 to 542 million years ago, of simple organisms with very primitive body plans, which had virtually no hard parts or exoskeletons, are some of the earliest known complex organisms (eg Fig.2) The South Australian museum houses a very good collection of these fossils.

Figure 2 Dickinsonia, Ediacaran fauna

Deposits of Ediacaran fauna have now been found in other parts of the world including China. The simplest and earliest known fossil life forms are again found in Australia and Canada and are formed from mats of cyanobacteria. There are fossil forms of these mats which are known as stromatolites (Fig. 3). The reason we can be fairly sure that the fossils are of stromatolites is that there are living examples of stromatolite mats in Western Australia at a place called Shark bay (Fig. 4) and recently some have been found living in the waters of an extinct volcano caldera called the Blue Lake at Mt Gambier in South Australia.

Many fossil stromatolites can be found throughout the world including Canada and Australia. The oldest known are from an area called the Warrawoona formation in a very hot and dry place, ironically called the North Pole, in Western Australia. They have been dated at 3.47 billion years old. This, and the methods used to date the rocks, is detailed in a book called "Life on a Young Planet" by Andrew Knoll [5]. Fossils of possible bacteria from times earlier than 3.47 billion years are the subject of fierce debate among paleontologists and suffice it to say the jury is still out on these specimens. Knoll develops these arguments in his book mentioned above.

Bacteria and archaea

It is not worth re-running the tape back though all the species exhaustively for this exercise (Dawkins has already done it in the Ancestor's Tale) but it should be enough to know that all species have arisen from previous species through an underbroken line of descent. We can go back through dinosaurs, amphibians, fish and worms to ultimately look at algae and other small multicelled creatures but these in turn have developed from single-celled organisms.

Ultimately the single-celled organisms which we know of as bacteria would have evolved from a precursor organism which is no longer present. Such is the pace at which bacteria evolve which can be seen by the rate that they adapt to entirely new food sources or other challenges. Therefore it is hard to determine which path their evolution may have taken. However, there is a very strong clue in the existence of another completely separate class of single celled organisms known as the archaea.

Archaea

Now classed as a separate family from the bacteria and from higher organisms, archaea have a set of unique chemical components which set them apart from bacteria. They are found in highly unusual environments and for that they are also known as extremophiles. For example, they can be found in huge numbers in soil, in hot springs, in highly acidic water and highly saline water. The features of these organisms that allow them to withstand these extreme conditions are not well understood but include, for example, the use of protein internal structures that resist decomposition by heat. A unique feature of archaea is the composition of their cell membranes which differ from bacterial cell walls in being composed of isoprenoid structures as opposed to the fatty acyl triglyceride structures of bacteria and eukaryotes. In the triglyceride molecules of most modern bacterial cells, fluidity of membranes is conferred by adding double bonds in the fatty acid structure. This is called a level of unsaturation and the greater the level of unsaturation the greater the membrane fluidity. In archaea the fluidity of their membranes is conferred by virtue of the branched chain structure of the isoprenoids. This will become important when we look at vesicle formation.

Oxygen and complex organisms

Stromatolites

The cyanobacteria have an extensive fossil record. The oldest known fossils, in fact, are cyanobacteria from Archaean rocks of western Australia, dated 3.5 billion years old. This may be somewhat surprising, since the oldest *rocks* are only a little older: 3.8 billion years old!

Figure 3 The cyano bacteria found in stromatolites are dated to a time prior to the advent of atmospheric oxygen.

Stromatolites of Shark Bay, WA

Figure 4, living stromatolites

Stromatolite or actually cyanobacterial photosynthesis would have contributed significantly to the production of atmospheric oxygen. The arrival of oxygen in turn allowed the development of multicellular organisms. Diffusion of oxygen through several cell layers made possible the addition of more than one layer of cells. Oxygen allowed organisms to develop a new way to derive more energy from their food. Respiration allowed organisms to make use of more energy from their carbohydrate food source by more efficiently oxidising the food completely to carbon dioxide. In doing so, animals were able to make the most efficient use of the oxidisation process. Up until the advent of oxygen in the atmosphere organisms could only use anaerobic (non-oxygen-dependent) metabolism to derive energy. This is akin to the process of fermentation in that it gave rise to a product such as lactic acid which contained three carbon atoms but could be utilized no further. Anaerobic metabolism of a glucose molecule could only reach a total of 140kcal/mole of energy from the process of glycolysis and in doing so its end product, pyruvic acid was a three carbon species which could convert easily into ethanol or into lactic acid.

Aerobic metabolism , (respiration) , took those three extra carbon atoms all the way through to carbon dioxide by a cyclic process known in modern biochemistry as the citric acid cycle, also known after its discoverer, Sir Hans Krebs, as the Krebs cycle. (Krebs recounted, in a lecture I attended, that his paper was rejected by several journals because the concept of a metabolic cycle was too radical at the time to be believed. It was of course ultimately accepted.). There are six carbon atoms altogether to be oxidized since six carbon glucose is split into two three carbon units during the anaerobic stage glycolysis process. This process allowed the amount of energy to be derived from one glucose molecule to reach 546 kcal/mole. The theoretical maximum amount of available energy from the process is 686 kcal/

mole so the glycolysis -citric acid cycle oxidation of glucose is 79.6% efficient (Fig. 5). Energy from a simple burning of glucose would only reach around 15-20% in most machines using combustion of carbon as the energy source.

Glycolysis and respiration via the citric acid cycle

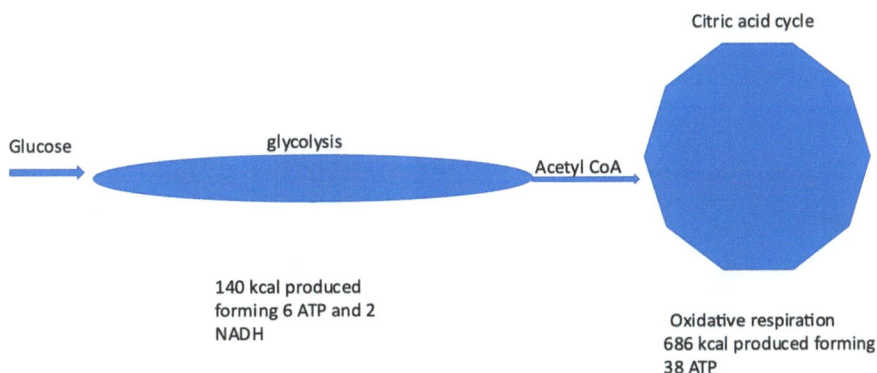

Citric acid cycle

Glucose

glycolysis

Acetyl CoA

140 kcal produced forming 6 ATP and 2 NADH

Oxidative respiration 686 kcal produced forming 38 ATP

Total reaction $C_6H_{12}O_6 + 6O_2 + 38 ADP + 38Pi \rightarrow 6CO_2 + 6H_2O + 38ATP$

Figure 5 Glycolysis and citric acid cycle

The energy released from this controlled stepwise oxidation process is stored in a molecule called adenosine triphosphate or ATP, a phosphate rich molecule with a sugar backbone and a nucleotide handle. ATP is the energy currency of organisms and it is used in the body to power everything from the synthesis of new cells to the energy required for muscle contraction. All of the above functions are necessary for the maintenance of life in complex organisms.

Early anaerobic life , however, made use of only the glycolysis part of the energy release. It was not until the advent of cyanobacteria which could trap atmospheric carbon dioxide and release oxygen as a waste product (photosynthesis) that the development of complex life, which could utilise oxygen in the aerobic steps, became possible. The release of oxygen into the atmosphere also created a huge amount of oxidized iron which can be found as a mineral, iron ore.

Last Universal Common Ancestor

Having outlined some of the latter steps leading to complex life and on up to mammals plants and ultimately humans, it is more challenging now to go back further and look at the possible steps leading to life in its simplest forms. If we go back and look at the single celled species of bacteria and archaea it becomes clear that both arose from a single precursor life form because both families have essentially the same genetic makeup. An organism that gave rise to both lineages has therefore been proposed and it is called the last universal common ancestor (LUCA).

What ingredients and conditions do we need to create LUCA?

1. Water
2. Carbon, nitrogen, oxygen, sulphur, phosphorus, iron.
3. Membrane vesicles or fatty acids that make them
4. Amino acids, and/or other organics
5. Template system for reproduction
6. Condensation conditions to allow formation of polymers such as polypeptides
7. An energy generating system, probably iron/sulphur or perhaps a proton gradient.

Where might we go looking for these ingredients and conditions?
Before doing that let us explore in chapter one the conditions that existed on a newly formed Earth and how anything could live on our newly formed planet?

Chapter 1

EARLY CHEMICAL DEVELOPMENTS ON A YOUNG PLANET

H ow did life on Earth arise? Did it arise here on Earth or did it come from outer space? There are many theories about how life could have arisen but it is impossible for certain to know how it happened as we cannot re-run the experiment on this planet. Because of this uncertainty there are those who prefer to send the problem off into space. If so, how did life arise out there in the cold of space --and how did it get here? The seeding of our planet by life forms or even by ready formed DNA is known as the theory of panspermia. One of its greatest proponents was Fred Hoyle, a famous physicist known for his theory of the formation of the elements of a higher atomic number than iron in the crucible of a supernova explosion. While this theory is well accepted to be correct and for which he should have received the Nobel prize, his theories about panspermia have not been well accepted. He was also famously wrong in arguing against the theory of the formation of the Universe by a big bang. His steady state theory [6] was overturned by this new Big Bang theory in the late 1960's. Hoyle coined the term 'big bang' because he thought he was poking fun at the new theory which has now become accepted by most cosmologists, including ironically the use of Hoyle's term 'big bang', for the phenomenon.

While some still hold on to the panspermia approach, which merely displaces the problem to another planet somewhere, I prefer to use the Ockham's razor approach of not proposing some complicated argument when a simpler one is available that makes sense. I propose to show in these pages that not only can we make a case for a set of simple steps leading to the first organism but that those steps are more likely to have occurred here on Earth than anywhere else in our corner of the Universe. In showing the logic of these steps I am also making the case for an understanding of the processes that have led to the magnificent burgeoning of all forms of life on Earth by the process of evolution. I have, in proposing the steps, taken the best available scientific evidence to hand. None of what i am proposing is in any way scientifically impossible. Indeed, all of the steps I am proposing have scientific evidence for them. Even further than that, there are some scientists who have proposed the existence of fanciful intermediate molecules eg. Szathmary and Maynard Smith who proposed coding coenzyme handles [7] to perform some of the early functions now ascribed to much more complex systems such as ribosomes in protein translation. I have tried to show that a set of steps using all of the known compounds and molecules that we know exist today, rather than invented ones, can be used to form the steps I am going to describe. There are many other books that have described some of these steps but I am trying to put forward a logical sequence using Ockham's razor to illustrate how molecules contributed to the existence of the first organism on Earth, LUCA.

The Last Universal Common Ancestor

1. There was a single organism scientists call the last universal common ancestor (LUCA) which must have been the precursor of all life existing on Earth today. How do we know this? Because all life that we have found on Earth has the same set of blueprints, the same code to make cellular components. The genetic code. So far, every living thing on Earth has been generated from another living thing and the process that makes that possible is the handing down of the code for making life, the genetic code (Fig. 6). The genetic code is a set of coding information which allows the translation from a DNA sequence (DNA, deoxyribonucleic acid) by a complex mechanism within every living cell into a series of gene products known as proteins. This is conducted by a number of enzymes which first copy the DNA encoding a particular gene into messenger RNA. This step is called **transcription**. The genetic information at that stage is known as genomic DNA because it has not yet been trimmed. The genomic messenger RNA, which contains extra sequences called introns is then modified by a number of enzymes into a sequence free of introns, the gene, that makes its way to a large complex of proteins and RNA called the ribosome where the triplet codons are **translated** into protein. Each amino acid of the protein is encoded by a unique triplet of DNA bases which are combinations of four DNA bases designated A,G,C&T for adenine, guanine, cytosine and thymine. There are 64 possible combinations of amino acids in these triplet bases which is more than enough to encode for the 20 amino acids used in making proteins.

However, the code shows U in place of T because that change occurs in RNA where Uracil replaces the Thymine found in DNA.

First base	Second base				Third base
	U	C	A	G	
U	Phe	Ser	Tyr	Cys	U
	Phe	Ser	Tyr	Cys	C
	Leu	Ser	Stop	Stop	A
	Leu	Ser	Stop	Trp	G
C	Leu	Pro	His	Arg	U
	Leu	Pro	His	Arg	C
	Leu	Pro	Gln	Arg	A
	Leu	Pro	Gln	Arg	G
A	Ile	Thr	Asn	Ser	U
	Ile	Thr	Asn	Ser	C
	Ile	Thr	Lys	Arg	A
	Met	Thr	Lys	Arg	G
G	Val	Ala	Asp	Gly	U
	Val	Ala	Asp	Gly	C
	Val	Ala	Glu	Gly	A
	Val	Ala	Glu	Gly	G

Figure 6, The Genetic Code

Note that the amino acids are written here as three letters for simplicity. Note that some amino acids are encoded by more than one triplet which introduces a level of redundancy into the code. In addition, though, the incidence of amino acid usage in proteins reflects the number of different codons for the amino acid. For example, there are six codons for leucine and it turns out to be the most frequently used amino acid in proteins.

The proteins which are made in the ribosome complexes conduct the running of all cell functions from building cell walls to converting food into energy and many thousands of other activities. The genetic code is a nearly universal code which is essentially the same in every organism, with some minor differences in primitive organisms. In humans the genome codes for about 30,000 gene products.

The fact that the code is universal is a very strong argument for there once having been a common ancestor of all life, the Last Universal Common Ancestor or LUCA. In fact, there must be an unbroken chain of organisms passing on their genetic material going all the way back to this organism because there is no other way that we know of that life can be transmitted from one organism to another other than via its DNA.

So, another possible definition for the last universal common ancestor could be that it was the first organism which relied completely on the genetic code, or something very much like it, for its existence. How long ago did the last universal common ancestor exist and what came before that?

There is an observation that a molecular clock can be worked out from the rate of accumulation of mutations.

> The notion of the existence of a so-called "molecular clock" was first attributed to Émile Zuckerkandl and Linus Pauling who, in 1962, noticed that the number of amino acid differences in hemoglobin between different lineages changes roughly linearly with time, as estimated from fossil evidence. They generalized this observation to assert that the rate of evolutionary change of any specified protein was approximately constant over time and over different lineages (known as the molecular clock hypothesis).

This is an observation of molecular biologists which uses the fact that very minor differences which occur in the amino acid sequences of the same proteins in different organisms made from the code give us a clue as to how long ago those organisms or species evolved from each other. The minor DNA differences occur at a roughly regular frequency based on the introduction of mutations from sources such as cosmic rays. The calculated time between mutations gives us a method which allows us to plot how, and from which, and how long ago, all species arose from their precursor species. For example one protein molecule used in all animal muscles and known as cytochrome C has been isolated and sequenced from a large number of different organisms. The cytochrome C of monkeys and cows is more similar than the cytochrome C of monkeys and fish. Such similarities and differences suggest that monkeys and cows are more closely related than are monkeys and fish. This then demonstrates that if we look at how each protein species, like cytochrome C from a whole range of organisms, differs in its protein sequence, these differences can be plotted on a 2D linear scale to give a molecular divergence tree of evolution (Fig. 7).

This tree is rather like those highly branched trees we see in books showing molecular clock information to show which species preceded which. The tree tells us how long ago each species diverged from its ancestors and as we go back far enough we come to the inescapable conclusion that there must have been a single organism at one time that eventually gave rise to every other species that has ever existed on Earth. There were almost certainly many other proto-organisms around at that early time but only one was ultimately successful at giving rise to everything else. LUCA, the last universal common ancestor. Interestingly though, the archaea seem to have diverged later than bacteria, but this just implies that the precursor of both bacteria and archaea would have had early features of both lineages. It does not preclude an archaea -like organism being the precursor and then diverging to fill both niches.

Phylogenetic Tree of Life

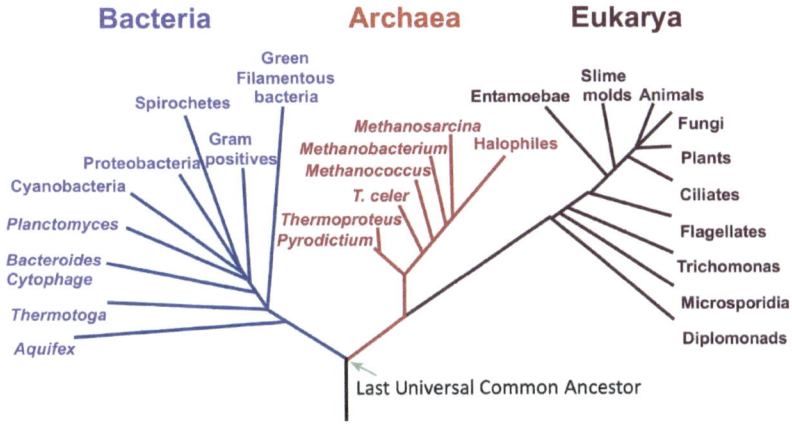

Figure 7 Phylogenetic tree of life

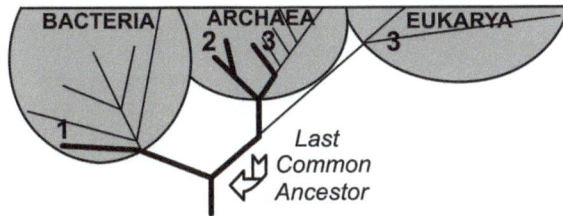

BACTERIA	T max (°C)		T max (°C)
1 Heat-Loving Bacteria			
Aquifex	95	Thermotoga	90
ARCHAEA			
2 Crenarchaeota			
Pyrolobus	113	Staphylothermus	98
Pyrodictium	110	Thermoproteus	97
Hyperthermus	108	Desulfurococcus	97
Pyrobaculum	104	Thermofilum	95
Gonococcus	103	Acidianus	95
Stetteria	102	Sulfophobococcus	95
Aeropyrum	100	Thermosphaera	90
Thermodiscus	98		
3 Euryarchaeota			
Methanopyrus	110	Ferroglobus	95
Pyrococcus	103	Archaeoglobus	92
Thermococcus	100	Methanococcus	91
Methanothermus	97		

Figure 8 Bacteria and archaea species

An examination of the archaea would be instructive at this point. As can be seen in the above table (Fig. 8) the archaea are named for the environments in which they are found or the activity they display.

Archaea are a separate form of life from bacteria and were so identified by microbiologist Carl Woese. They are also called extremophiles as they are found in environments which most other organisms cannot tolerate. High salinity, high temperature, low pH (high acidity) are all environments in which these extremophiles are able to survive. Their makeup is such that proteins are able to withstand these extremes without denaturing. They also have membrane structures which are different from modern biochemical systems in that they are made of isoprenoid structures which have side branched chains. Isoprenoids confer fluidity properties to their membranes while modern organisms use fatty acid unsaturation (double bonds) to achieve the same result.

So there was a single precursor organism, LUCA (Fig. 9), but how did it arise? I propose in the rest of this book to suggest possible ways in which this occurred. There is no real way to know this for sure since the experiment cannot be re-run without being able to reproduce exactly the prevailing conditions at the time and these we can only guess at. Nevertheless, having followed the literature for many years, I intend to show that there is enough evidence to make a plausible case for the sequence of events that might in all probability have occurred. I make no apology for the fact that the sequence is my interpretation of the available data and may differ from many other proposals. It is also a fact that much of what occurred would have been at the molecular level and involves chemical processes which it is difficult to depict. Therefore, there is going to be some understanding of chemistry needed to fully appreciate what went on. I will try to explain this as we go along.

Steps leading up to the last universal common ancestor

When was the Earth first able to support life?

Delivery of water and organics to the cooling Earth

Formation of vesicles in fire and ice

Order from chaos. Molecules capable of self-assembly (the role of entropy)

Clay templates for polymer formation-peptides?

Self reproduction by limited peptides

Primitive enzymes from ancestral domains

Synthesis of more complicated molecules--> RNA

Production of Ribozymes, the RNA world and/or the RNA-peptide world

Development of the genetic code

Voila LUCA.

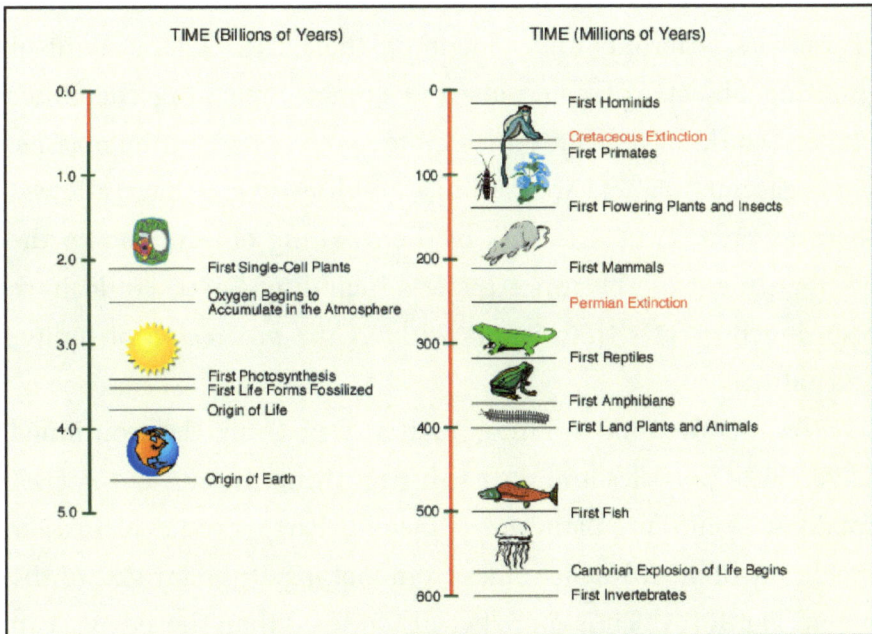

Figure 9, When was the Earth first able to support life?

The Earth formed about 4.6 billion years ago (Fig. 9), with the first fossilized life evident about 7-800 million years later at 3.8 billion years ago and the first photosynthesizing life about 1 billion years after that. Oxygen did not begin accumulating in the atmosphere until about 2.5 billion years ago. The solar system is believed to have condensed from a huge dense cloud of dust which encircled the sun. The disk of dust and gasses, while spinning, gradually condensed by gravity into discreet objects which became the sun and its orbiting planets. Each planet would have initially been molten and then cooled to form the rocky planets in the case of the four inner planets, Mercury, Venus, Earth and Mars. The outer, gaseous planets would also have condensed but were left with massive dense cores that could retain huge atmospheres of gases. Some disks did not condense into planets and became the asteroid belt. During the cooling period, after the formation of the rocky core of the Earth, there was a time of massive bombardment of meteors, asteroids and comets on the Earth. The amount of infalling objects, of asteroids and comets, reaching the inner planets would have been protected to some extent by the massive outer gaseous planets whose huge gravities would have allowed them to vacuum up the bulk of the infalling objects and so the bombardment of the inner planets including Earth could have been much greater than it was without the protection of Jupiter and Saturn.

Presumably this infalling matter was from the continued accretion of particles and dust still circulating the sun but not yet condensed onto any planet or planetoid. Part of the evidence for this heavy bombardment comes from looking at the far side of the moon (Fig. 10) which is more heavily cratered than the side we can see facing Earth.

Figure 10, the far side of the moon

It would have taken the same number of hits as the Earth per surface area during the bombardment period, the only difference now being that the Earth has undergone erosion and biological modifications which have obliterated most of the original cratering. A few large craters are still evident on Earth though, including some which have only recently been recognised as such. One example is the asteroid impact which formed Lake Acraman in South Australia some 580 million years ago leaving a nearly circular impact scar that now contains a lake (Fig. 11) [8].

Figure 11 Lake Acraman, site of the original impact crater

There are many other examples of such impact craters but the vast majority of them have been obliterated with time and erosion. The notable impact crater in the Yucatan peninsula of central America is now thought to be the impact which caused the demise of the dinosaurs and in doing so allowed the rise of mammals to occur. Had that impact not occurred to wipe out the well-established dinosaurs, which had been on Earth for 200 million years, it is almost certain that we humans and most mammal species would not be here today.

The early period of heavy bombardment involved energies capable of vapourising anything volatile on the Earth's surface which includes water, most organic material and other gases. These would have been driven off into space under this continued highly energetic pummeling of Earth and moon from space.

Delivery of water and organic compounds to the cooling Earth

The figure below (Fig. 12) shows the energies involved and shows that the period of heavy bombardment only began to ease off at around 3.8 billion years ago. It was not until this point, at which the energies of the impacts had reduced enough, that some of the volatile material could remain on Earth. Conditions of milder impact energies would have allowed the accumulation of these materials on the surface. Under this reasoning all of the water on the Earth would have to have come from asteroid and comet impacts since, prior to that point, impacts would have been energetic enough that water would have been blown off into space. In addition, most if not all, of the organic material on Earth had to have come from these extraterrestrial sources as well since carbonaceous material would also have been volatile and blown off into space prior to the 3.8 billion year inflection point.

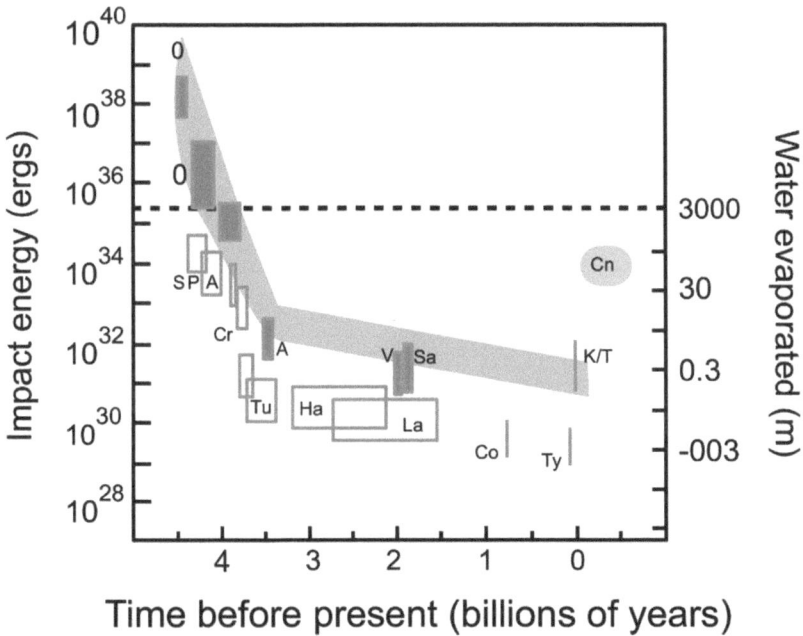

Figure 12 Impact energies vs time ref [9,10]

There is still some argument about the origin of water among scientists who show that the ratio of isotopes in the water of several comets is not the same as the isotope ratio in Earth's waters. A finding from the 2014 Philae lander shows that the isotope ratio on the comet 67P/Churyumov-Gerasimenko is still an unfavourable match for this theory. Philae lander on comet 67P (Fig. 13)

The ratio of deuterium to hydrogen is three times too high to account for this comet being from among those that gave their water to the developing Earth. However, it is believed that this is a Jovian comet whereas the water on Earth is likely to have come from the asteroid belt, much closer in. It is now thought that at least some asteroids have their own oceans.?? At least one comet has been shown to have the same isotope ratio as the water here on Earth. In conclusion, water and much of the organic material on Earth has come from sources off the planet and probably from asteroids and comets in the nearer asteroid belt. As I have indicated, this is still an area of continued research and more data is needed to confirm this hypothesis, but it remains the most plausible at this time.

The early period of heavy bombardment involved tremendous energies at fairly high temperatures capable of vapourising anything volatile on the Earth's surface which includes water, most organic material and all gases. These would have been driven off into space under this continued pummeling from space.

The figure shows the energies involved and shows that the period of heavy bombardment only began to ease off at around 3.8 billion years ago. Source [Ref 9], Comets and the origin and evolution of life, 1st ed p199

It was not until this inflection point, at which the energies of the impacts had reduced enough, at about 3.8 billion years ago for some of the volatile material to remain on Earth. Conditions of milder impact energies would have allowed the accumulation of these materials on the cooling Earth's surface. Under this reasoning all of the water on the Earth would have to have come from asteroid and comet impacts. Prior to that time heavy bombardment impacts would have been energetic enough that all surface water would have been blown off into space. In addition, most if not all of the organic material on Earth had to have come from these extraterrestrial sources as well since it would also have been volatile and blown off into space prior to the inflection point.

There is still some argument about the origin of water among scientists who show that the ratio of isotopes in the water of several comets is not the same as the isotope ratio in Earth's waters. A finding from the 2014 Philae lander (pictured, Fig.13) shows that the isotope ratio on the comet on which it landed, 67P/Churyumov-Gerasimenko, is still an unfavourable match for this theory.

Figure 13, Philae lander on the comet 67P

An explanation for the observed hydrogen isotope ratio

There may be another explanation for the high D/H ratio on comet 67P. If it was formed a long time ago, as many comets were, then the surface would have been exposed to the hard vacuum of space over that long time of perhaps millions of years. As ice on the surface tends to sublime in a vacuum, going directly from solid ice to vapour, there would have been a gradual concentration of the heavier D isotope of hydrogen in water (heavy water) because it has a higher latent heat of evaporation. Meaning it takes more energy for the heavy water D_2O to sublime in a vacuum than H_2O. This would gradually tend to concentrate D_2O isotope on the surface over a long time. I suspect that measurements were very likely taken on the surface where the exposure to vacuum was greatest, not deeper down. A way to test this would be for future moon missions to measure the D isotope in the water there as it would also have been exposed to the hard vacuum of space while water on the Earth, which was presumably deposited by comet impacts at the same time, has not had the same exposure. I would predict the moon ice has a higher D/H ratio than water on the Earth. There will be quite a variation in these measurements depending on how long ago the water arrived on the moon. The same measurements could be done on another comet but would have to include drilling down deep into the ice to obtain samples not exposed to vacuum as well as taking surface measurements. As I have indicated, this is still an area of continued research and more data is needed to confirm this or any other hypothesis but it remains the most plausible to me at this time.

It is very unlikely that there may have been any native water as the early planet cooled but the majority eventually was derived from extraterrestrial sources, in particular, asteroids and comets. A recent study has shown that water may have been on Earth from a time much closer to the beginnings of Earth, but I think this needs some checking. Studies of the Moon's craters and their ages suggest there was a similar early high impact, high energy period. The moon's gravity was too weak to retain much of the water or an atmosphere. Recently, however, water ice has been discovered on the moon deep in some craters which do not get much exposure to sunlight.

During the period of heavy bombardment on Earth, prior to 3.8 billion years ago, the collision energies would have been so high that any surface water would have evaporated away back into space but as energy and impact rates declined after the 3.8 billion year inflexion point, water and other volatiles would have gradually begun to be retained. There is plenty of water in the solar system eg. Saturn's rings are made up almost entirely of ice, and it is thought that many of the objects in the Oort cloud, on the outer regions of the solar system, are icy bodies. These bodies are periodically knocked out of their orbits by the sun to become comets which orbit the sun and therefore have the potential to impact other solar system bodies. Similarly, water has been detected on some of the asteroids closer in to the Earth. Moons such as Jupiter's Europa and Saturn's Enceladus have a lot of ice on their surfaces and possibly hidden oceans.

Stanley Miller's experiments

An experiment that literally electrified the scientific world and virtually began the science of prebiotic chemistry was conducted by Stanley Miller in the laboratory of Harold Urey in the 1950's. Miller and Urey surmised that the early Earth atmosphere probably contained the following gases: methane, hydrogen, water vapour and ammonia. These he subjected to an electric spark over many days while circulating water through the apparatus. The remarkable result was that several amino acids were produced in this process (Fig. 14).

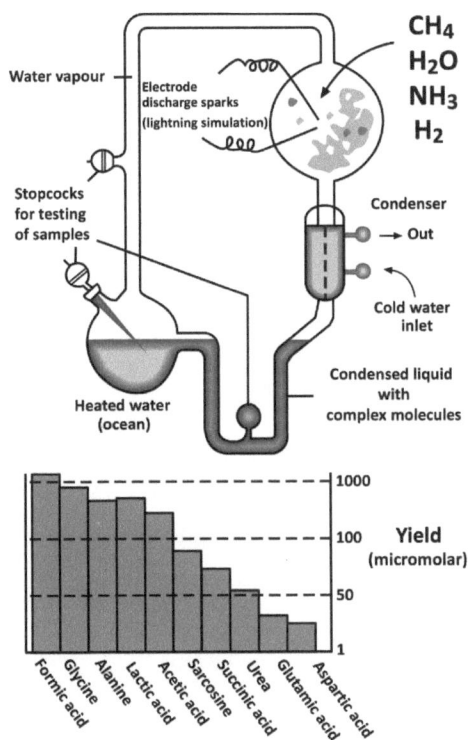

Figure 14, the Miller /Urey experiment

In the resultant brown coloured water Miller was able to identify the formation of <u>amino acids</u>, such as <u>glycine</u>, α- and β-<u>alanine</u>, using paper chromatography. He also detected aspartic acid and gamma-amino butyric acid. This experiment has become a classic textbook definition of the scientific basis of origin of life and is regarded by some as the first definitive experimental evidence of the Oparin-Haldane's "primordial soup" theory (Fig. 15).

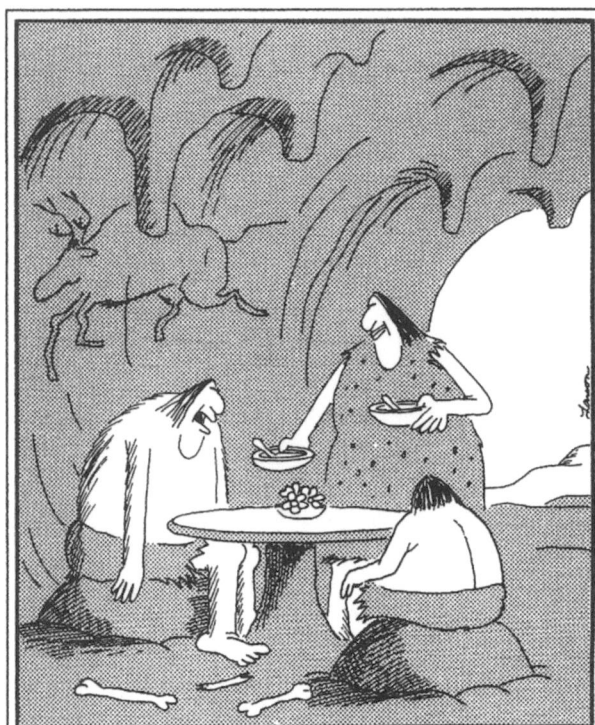

"Primordial soup again?"

Figure 15, Primordial soup again?

However, the estimated composition of the Earth atmosphere in those early times has been questioned and in particular there would not be as much hydrogen in the atmosphere as proposed in the Miller Urey experiment and no carbon dioxide was included either. Nevertheless, the experiment got people's attention and ignited a whole field of prebiotic chemistry.

The ingredients: Delivery of water and organics to the cooling Earth

Some scientists [11] postulate mechanisms for the formation of biologically important molecules from simple starting materials such as the formation of adenine from hydrogen cyanide (Fig.16).

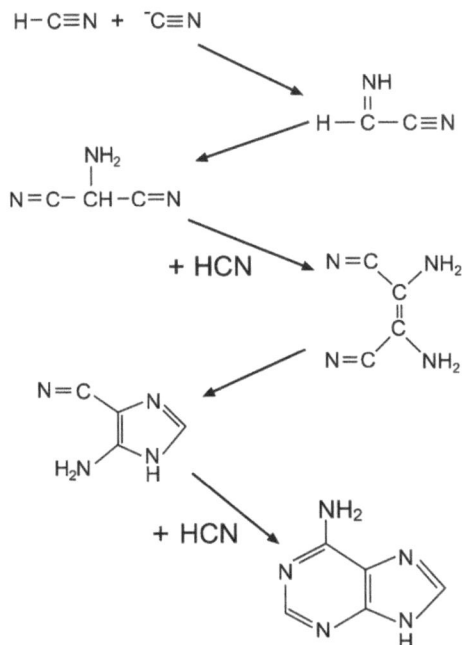

Figure 16, prebiotic synthesis of adenine from hydrogen cyanide

Similarly the Strecker synthesis (Fig.17), [12] has been proposed for amino acids by the **reaction** of an aldehyde with ammonium chloride in the presence of potassium cyanide.

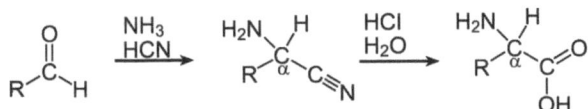

Wickipedia

Figure 17, Strecker synthesis of amino acids

A route for the synthesis of sugars from formaldehyde called the formose reaction has also been proposed (Fig. 18), [13].

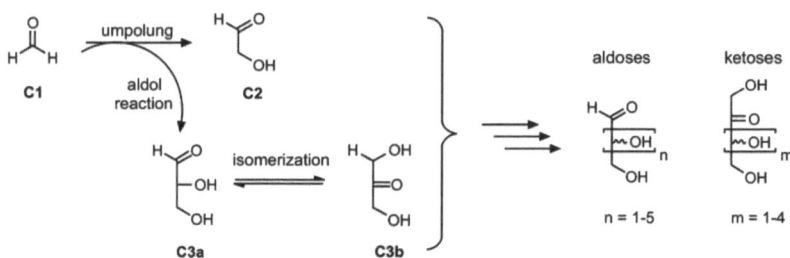

Figure 18, Formose synthesis of sugars

However:

In a collection of articles under the title "Comets and the origin of life, volumes 1 and 2" [1,2] that brings together a lot of research in the area, data on this period of Earth's history can be found.

As we were reminded only recently (2013) with the arrival of a body which crashed to Earth near the Russian town of Chelyabinsk causing quite a bit of damage, the Earth continues to receive much material from space.

The **Chelyabinsk meteor** was a superbolide that entered Earth's atmosphere over Russia on 15 February 2013 at about 09:20 YEKT (03:20 UTC). It was caused by an approximately 20 m (66 ft) near-Earth **asteroid** with a speed of 19.16 ± 0.15 kilometres per second (60,000–69,000 km/h or 40,000–42,900 mph). (Wikipedia)

That asteroid was 20 metres in diameter but at the same time a near Earth object several times larger was observed passing by between the moon and Earth. A collision on Earth from that object would have been very much more catastrophic. It is estimated that we receive approx. 9 tonnes of material annually falling from space. Another estimate is 40,000 tonnes which arrive at the atmosphere but much of it burns up. Some of it is in the form of chondrites, small black glass-like nodules made of organosilicates, which contain a range of organic compounds. They typically contain up to 6% carbon in various forms all of which can eventually find its way into other organic compounds some of which become incorporated into living systems. Chondrites can be found in places such as deserts and ice and snow sheets in Antarctica where the black material is easier to spot. A magnet dragged through your roof gutters may also be a way to find ferrometallic nodules which are continually falling from the constant meteorites raining down on the Earth.

Analysis of the dust from comet Halley also contains some very interesting compounds including purines and pyrimidines which are components of DNA and RNA.

The synthesis of building blocks such as amino acids, nucleosides and sugars have all been envisaged as arising spontaneously by various mechanisms. Sugars, amino acids and nucleoside bases can be derived in proposed mechanisms including the Strecker synthesis of amino acids from hydrogen cyanide and of other syntheses from precursor formaldehyde and formaldimine. The proponents of these reactions are criticized because they are assuming pure reactants and the amounts of product are too small to be significant.

A cometary origin of these raw materials becomes more likely. Carl Sagan and coworkers identified some of these precursor molecules in space. It is possible that some of this chemistry to make amino acids, sugars and nucleosides went on in the ice of comets. While slow, because of the icy temperature, the reactions could still have occurred to produce significant amounts of product over the eons that the ice of comets spends in space.

On early Earth, Ehrenfreund at al (2002) [9] estimated possible terrestrial sources of organic materials were the following;

- UV photolysis ($3x10^8$ kg/yr),
- electric discharge ($3x10^7$ kg/yr), (As in the Miller Urey experiment)
- hydrothermal vents($4x10^2$ kg/yr)
- shocks from extraterrestrial impacts ($1x10^8$ kg/yr)

while extraterrestrial sources were:

- interplanetary dust ($2x10^8$ kg/yr) and
- comets ($1x10^{11}$ kg/yr).

Thus comets, meteorites and asteroids were estimated to provide 1000 times more organic material than any other source. This would have included organic molecules such as some amino acids and some purines and pyrimidines. These are the basic ingredients needed to form proteins from amino acids and to form DNA and RNA from purines and pyrimidines. It is suggested that the majority of organic material on the Earth today, including these essential for life ingredients, has an extraterrestrial source! Where did all this raw material come from and how much was there?

The total water mass on Earth is about $1.4×10^{21}$ kg.[3]

The total mass of carbon in living systems is estimated to be $7.7x10^{17}$ g [14] while the total carbon mass on Earth is estimated to be $7.7x10^{22}$ g. This latter figure would include rock as carbonates including limestone, dolomite and other minerals. Carbonates are essentially trapped carbon dioxide as minerals. The trapping was done by coral formation and similar processes. So the original form of carbon is not known but may have been atmospheric methane.

See inset below.

The source of carbon on Earth is still a matter of debate but:

Presenter: George Shaw, *Union College*

When: October 24, 2006 2:30PM PDT

It is widely agreed that carbon first arrived on Earth in a reduced form, as found in almost all meteorites, and was abiotic in origin. For more than thirty years, the prevailing view has been that the carbon in Earth's early atmosphere (and near surface environment) was virtually all in the form of carbon dioxide, the oxidized chemical state found in volcanic gases that are thought to be the source of atmospheric carbon compounds resulting from degassing of Earth's interior. An early reduced carbon reservoir at/near Earth's surface follows directly from early degassing, under reducing conditions, of the original (and/or hydrogenated) meteoritic carbon compounds. The largely methane atmosphere so produced is short lived, but the photochemical products accumulate in the ocean and are continuously recycled into the atmosphere as methane by low temperature hydrothermal activity. This model provides a suitable source of the early (methane) enhanced greenhouse effect.

Recent observations suggest that space is full of fatty type molecules. An example is the Murchison Meteorite which came to Earth in 1969 and was discovered in regional Victoria, Australia. It has a noticeable turpentine odour and is rich in organic compounds. Mass spectrometry revealed the presence of over 14000 unique organic compounds in the meteorite. Miller points out that eighteen amino acids have been identified in the Murchison meteorite (Fig. 19), a type II carbonaceous chondrite, of which six occur in proteins.

All of the amino acids found in the Murchison meteorite have been found among the electric discharge products of Miller's experiment.

Figure 19, the Murchison meteorite

Conclusion: much of the raw materials, water, organic compounds, fatty molecules, for the formation of life arrived from outer space associated with comets and asteroids.

Chapter 2

ORDER FROM CHAOS

FORMATION OF VESICLES AND THEIR CONTENTS

Solvent extracts of the organic material from the Murchison meteorite was shown to be fatty enough to be capable of forming lipid vesicles. Therefore, the meteorite contained fatty acids and other lipid compounds which were amphiphilic which means that they had one end which was hydrophilic (polar), allowing interaction with water and the other end was fatty (non-polar) and repelled water. The diagram shows the way these molecules aggregate to form what are called lipid bilayers where all of the polar heads are aligned to interact with the water medium while the non-polar tails all align together and interact with another array of lipid tails to form a bilayer. This bilayer rounds up to form a spherical vesicle which will contain material dissolved in the inner water medium and at the same time the exterior water medium is part of the larger medium in which the vesicles exist and float about. If these media are different it can set up a gradient across the vesicle membrane either a chemical one or an electrolytic one. One of the simplest is a salt gradient where the inner and outer salt concentrations are different. This sets up an osmotic pressure difference which can either swell or shrink the vesicle making the vesicle leaky and able to absorb chemicals. This can then result in a new content of perhaps amino acids or other materials. In this way vesicle contents can result in many new combinations of chemicals and organic substances. Vesicles can also fuse together or split apart and mixing of contents is a result. Vesicle formation is the first example of order from chaos where self-assembly of molecules forms a structure conducive to the first steps in the formation of LUCA (Fig. 20).

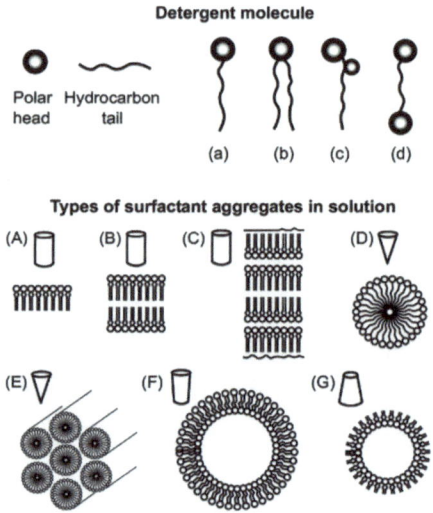

How might fatty acids have formed on the early Earth? Some scientists have proposed that **hydrothermal vents** may have been sites where prebiotically important molecules, including fatty acids, were formed. They propose a theoretical scenario in which fatty acids are formed along the face of a **geyser.** Research has shown that some minerals can catalyze the stepwise formation of hydrocarbon tails of fatty acids from hydrogen and carbon monoxide gases -- gases that may have been released from hydrothermal vents. Fatty acids of various lengths are eventually released into the surrounding water.

Figure 20, formation of vesicles from fatty acids

Now that we have established a probable source of ingredients how did this begin to take shape as life?

Scientists have been thinking about this question for a long time. Freeman Dyson in his book Origins of Life [15], lists 3 different theories: each addresses the question differently of how did both metabolic processes and self reproduction processes evolve? Most more recent theories are variations on one or more of these themes.

1. Oparin 1924, Haldane 1929 both considered that cells came first, then enzymes, peptides followed by reproductive molecules (they preceded Watson/Crick 1953 and didn't know about DNA)
2. Eigen/Orgel 1981, RNA first, protein, enzymes then cells.
3. Cairns-Smith 1982, first clay, followed by enzymes, then cells followed by genes.

All but the first of these theories do not place the formation of cells as a priority in the development of life.

How much building material for life is there on Earth? The total mass of carbon in living systems is estimated to be $7.7x10^{17}$ g (that is about 800 billion tonnes) while the total carbon mass on Earth is estimated to be $7.7x10^{22}$ g (Biogeochemistry, Schlesinger 1997). This latter figure would include coal and oil deposits and rock as carbonates including limestone (Calcium carbonate) and dolomite (Calcium-magnesium carbonate). Oil and coal are proposed to have been made from decaying organic matter from early periods of life on Earth but an alternative theory proposed by Tom Gold (The Deep, Hot Biosphere 2001) proposes that they are minerals deposited on the early Earth. Most geologists would challenge that view but the arguments put forward in Gold's book are not easily dismissed, particularly relating to the bacteria found in soil at great depths and near oil deposits. Carbonates are essentially made from atmospheric carbon dioxide which has been trapped as these carbonate minerals either by chemical processes or by biological processes such as the formation of coral reefs in shallow seas. Sea shells are also made of calcium carbonate. Much of the trapping was done by molluscs, forams and corals. An early coral like formation that formed huge reefs in southern Australia was an organism known as Archaeocyatha (Fig. 21). A walk along parts of the Flinders Ranges in South Australia allows the hiker to find in places that almost every rock is limestone formed of fossilised Archaeocyatha.

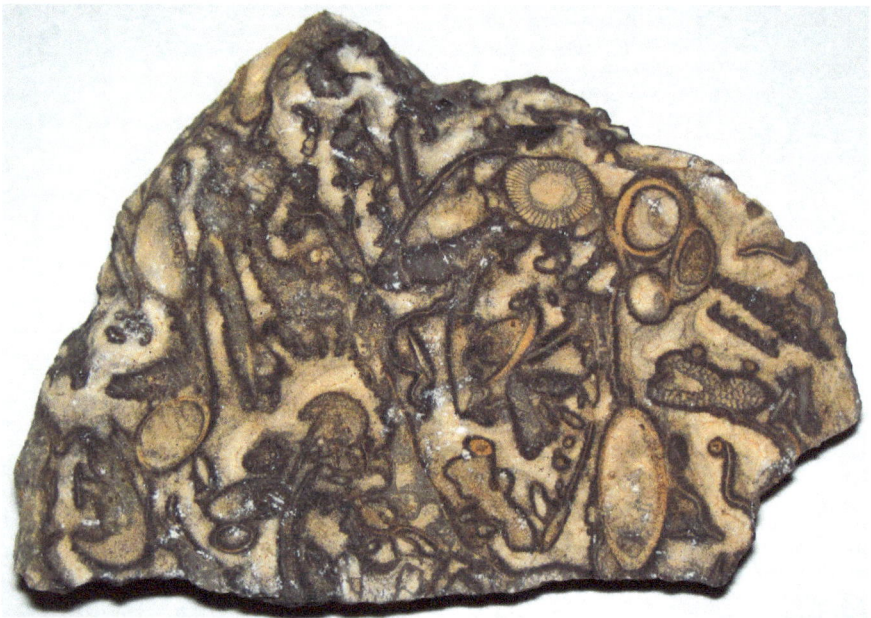

Figure 21, Fossils of Archaeocyatha, a prehistoric coral.

Marble, which is metamorphosed limestone may contain fossil remnants of its origin from various marine organisms.

One conclusion is that carbon existed as carbon dioxide in the atmosphere probably from earliest times by the outgassing of volcanoes. Some volcanoes still expel carbon dioxide gasses such as one in the Cameroon which killed a lot of people and cattle by suffocation.

On 21 August 1986, a limnic eruption at Lake Nyos in northwestern Cameroon killed 1,746 people and 3,500 livestock.

The eruption triggered the sudden release of about 100,000–300,000 tons (1.6 million tons, according to some sources) of carbon dioxide (CO2). The gas cloud initially rose at nearly 100 kilometres per hour (62 mph) and then, being heavier than air, descended onto nearby villages, displacing all the air and suffocating people and livestock within 25 kilometres (16 mi) of the lake. Source Wikipedia

If there was a large amount of atmospheric CO_2 it would have trapped a great deal of solar radiation which would have raised global temperatures but this would have been in a period when the sun was dimmer. The lower solar radiation level would have made the global temperature cooler and the planet would have been warmed by the greenhouse effect enabled by atmospheric CO_2.

The total water mass on Earth is about 1.4×10^{21} kg.

To put these enormous figures into perspective that is enough water to fill 4/5 of the Earth's surface to a depth of up to 3 km, on average.

Conclusion: The first steps towards the formation of vesicles capable of developing into cells is described. Other ingredients such as water and atmospheric gasses are described.

THE PREBIOTIC ERA AND MOLECULAR PALAEONTOLOGY

The prebiotic era is a period preceding Darwinian evolution where the development of chemical processes occurred. It is the first part of the period leading to the formation of LUCA. A question is whether events in this era can be termed 'chemical evolution'? Some studies of the era include those by P.G. Higgs, Chemical evolution and the evolutionary definition of life [16]. To quote Higgs, "Chemical evolution is an important stage on the pathway to life, between the stage of "just chemistry" and the stage of full biological evolution". Also those by M. Tessera, Is pre-Darwinian evolution plausible? [17] In this detailed paper Tessera discusses the advantages and disadvantages of several prebiotic evolution models. A paper by D. Kunnev, Origin of Life: the point of no return [18]. Kunnev suggests that the initiation of Darwinian evolution constitutes a point of no return after which life begins. While the above theoretical studies indicate possible early processes, the area of molecular palaeontology offers some, perhaps more solid clues to the era.

Palaeontology is a remarkable science in how much information can be obtained about an organism from a few fragments of shells and bones. Its practitioners can often tell how old, how big, what it ate, what it looked like and much more, sometimes from just a jawbone fragment. In this way some remarkable stories have been put together of our Earth's past inhabitants. One of the most spectacular palaeontology stories is of the Burgess shale deposits in Yoho National Park, Canada in which fossils of the so-called pre-Cambrian explosion of life forms were preserved. This story was eloquently told in Stephen Jay Gould's "Wonderful life".

An enduring theme from that work which has importance for this story is the occurrence of periodic extinctions throughout geologic time and the implications for the surviving species. It is Gould's premise that the species that survive such extinctions do so through no apparent fitness for survival. It seems to be just pure luck that they were in the right niche at the right time. They might have been able to withstand a greater period of darkness or cold than other species for example. The surviving species then go on to evolve and fill the available ecological niches vacated by the recent extinction based on a new, but restricted to the precursor's, set of body plans and perhaps a restricted (or refined) set of biochemical pathways.

Molecular palaeontology is just as fascinating as the more familiar variety because potentially it can tell us about the process of the evolution of life on Earth. Before there was life there must have been a period of evolution or 'experimentation' of individual molecules, which could somehow replicate or be copied and this is called the 'prebiotic era'. [Not to be confused with the emerging field of nutrition called Pre- or Pro biotics which seems to refer to food that can replace or nurture resident gut flora]. We can never know for sure what this prebiotic era was like but there is certainly a large amount of cerebral energy expended trying to guess. Some of the early pioneers in this area turn out to have had some of the most original ideas on the subject and I will explore some of their theories.

Early observations that led to the RNA world hypothesis

Molecular palaeontology, or the study of molecular fossils of the prebiotic era (as opposed to the study of ancient molecules in fossils which is also a fascinating area of study), first came to my attention in a paper by Dutch researchers Visser and Kellogg (**Visser**, C.M., **Kellogg**, R.M. Biotin. Its place in evolution. J Mol Evol 11, 171–187 (1978)) [19,21]. These researchers studied model compounds that were designed to emulate the chemical reactivity of biotin. Biotin is a vitamin (vitamin H) involved in binding carbon dioxide as bicarbonate and transferring it to a range of acceptor molecules in cells. Biotin is found in the central active site region of several important metabolic enzymes called carboxylases that all bind carbon dioxide, usually as bicarbonate ion, and attach it to an acceptor molecule to generate a new compound containing an extra carboxyl (COOH) group. Biotin is what is known as a prosthetic group because it was not made as part of the original protein or enzyme sequence but attached later in a post-translational modification to the protein. The enzyme protein as translated from the messenger RNA code is made exclusively of amino acids. Biotin, which is not an amino acid, is attached to a lysine amino acid in the apocarboxylase protein molecule using another enzyme called biotin holocarboxalase synthetase, which has evolved specifically for this purpose. This is very much like the addition of a false limb so it is called a prosthetic group. Apo- and holo- are prefixes describing the underivatized and derivatised enzyme molecule respectively.

Biotin, which is contained in, for example, the mitochondrial enzyme pyruvate carboxylase (PC), catalyses the following two step reaction, which is essentially the fixation of carbon dioxide into a small metabolic intermediate, pyruvate, which is central to all cellular metabolism. Biotin is the link between the two steps, holding the CO2 temporarily between the steps. Pyruvate is a three-carbon compound while the product, oxaloacetic acid, is a four-carbon compound, which goes on to be involved in further reactions which contribute to the oxidative metabolism of sugars.

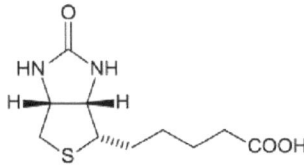

Structure

Step 1: HCO_{3-} (bicarbonate)+ Adenosine triphosphate (ATP) + biotin-(PC) → (PC)-Biotin-CO2 + adenosine diphosphate (ADP) + inorganic phosphate

Step 2: Pyruvate + (PC)-biotin-CO2 → Oxaloacetic acid + Biotin-(PC)

Visser and Kellogg found from their model compound studies that biotin has no innate chemical reactivity of its own and was probably originally another substrate molecule metabolised in the cell just like many other molecules. Attaching it actually makes it more efficient as the enzyme does not have to find it in the cytoplasm and its presence makes the reaction more efficient due to proximity. The lack of innate reactivity of biotin was very unlike the innate chemical reactivities found in other molecules which bind to enzymes to assist catalysis. These include the nucleotide cofactors such as adenosine triphosphate (ATP) and acetyl coenzyme A and some ribonucleic acids (eg transfer RNA and the ribozymes). Cofactors are compounds employed by enzymes as carriers within the cell of sources of either energy, hydrogen, methyl groups, or other commonly used groups. ATP has a highly energetic bond between its last two phosphate groups, which allows cells to use this molecule as an energy currency. It is formed by the controlled oxidation of sugars (which we know as respiration) and then moves about the cell. The energetic bond can be readily split in enzyme catalysed reactions and the energy transferred to other molecules for a variety of cell functions such as forming new molecules, powering reactions requiring energy and most importantly making muscles move. Acetyl Coenzyme A has a reactive or energy rich bond too, the thioester mentioned in De Duve's theory (see " Vital Dust" by C De Duve [21]) that allows easy transfer of acetyl (CH3-CO-) groups in a variety of cell reactions. Ribonucleotides are even more interesting in that they can fold in a number of ways, can contain genetic information and can catalyse reactions themselves. They concluded that biotin became incorporated into carboxylases to provide catalytic advantages of approximation by reduced degrees of freedom (because one end of the molecule is physically attached to the enzyme and cannot drift away. This is merely an efficient extension of the way all enzymes work in placing specific compounds close together (approximation) to facilitate their reaction together to form a new compound. The covalent attachment of one of the reactants, in this case biotin, reduces the need to obtain it from the cell cytoplasm thus improving the efficiency of the reaction.

The difference between this kind of catalytic advantage and the innate chemical reactivities of nucleotides gave Visser and Kellogg the idea that nucleotides may be survivors, or molecular fossils, of a period in the prebiotic era called the 'RNA world'. W. Gilbert developed this idea into a theory, (which can never directly be proven) which is now accepted (with some notable dissentions) as a very plausible explanation for part of the prebiotic stage in the development of life on Earth. I will return to this concept after we have considered what the minimum requirements for a living organism might be.

Conclusion: Exploring the role for early nucleotides in development of nucleotide compounds with innate reactivity and by co-association protection of peptides in an early prebiotic stage. Suggestion of the RNA world which later develops into a full theory.

MOLECULES CAPABLE OF SELF-ASSEMBLY: MORE ORDER FROM CHAOS.

At several stages in the prebiotic era there must have been molecular self-assembly going on. If we take the brown liquid that Stanley Miller formed in his classic experiment (see box above) and let it stand for a long time, there will never be more complex ordered molecules such as polymers arising from it. The reason is that according to the laws of thermodynamics chaos always increases or entropy will always cause any system to tend towards disorder. The classic analogy is the unlikelihood of a group of monkeys assembling by chance a Boeing 747 aircraft from material in a junk yard. Order from chaos just doesn't happen without some other driving force. That is if there is an extra set of circumstances where the chaos can drive self assembly. One of the greatest challenges facing those who speculate about the origin of life on Earth is how biological systems or prebiotic systems overcome the universal tendency towards disorder. Disorder is known scientifically as entropy and it always tends towards more disorder or higher levels of entropy. The second law of thermodynamics says that in any system entropy always increases with time. However, biological systems clearly do not <u>appear</u> to obey that rule. We can see that most inanimate and even living systems around us do obey that rule where order is gradually replaced by chaos. Erosion gradually levels mountains, weeds invade orderly gardens unless constantly removed, structures such as bridges require constant attention to prevent rust from damaging them, buildings will decay if left unattended.

One such example of order from chaos is the common observation we may have of a familiar self-assembly process, crystal formation. Allow a solution of common salt to dry out slowly enough and a lot of small crystals will form. The more slowly the process is allowed to take the larger the crystals that form. The process is counter-intuitive to thermodynamic principles in that the <u>self-assembly</u> of molecules seems to be defying the tendency towards chaos or entropy. But in fact, what is driving the process is the extra chaos imparted to the solvent, in this case water, in going from keeping individual molecules of salt in solution, which relies on a partly structured or entrapped water, to becoming free, more chaotic water. Structured water is a form of hydrogen bonded water (Fig. 22) that is on the way to forming the regular structures that occur in ice.

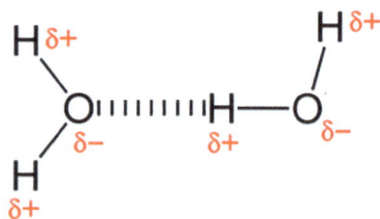

Figure 22, Hydrogen bonded water showing the residual charges on each atom which allow the bonding to occur.

Because of its open structure caused by this hydrogen bonding, which makes it less dense, ice floats in water. It is this structured water which also forms around molecules as a cage (clathrate) such as around sodium ions and chloride ions to keep them in solution so when the sodium and chlorine atoms are arranged in a crystal lattice which excludes water there is a release of water from its structured form that keeps the sodium ions and chloride ions in solution. The water takes on a more chaotic form, less hydrogen bonded, and this releases entropic energy. The release of energies in each molecule of sodium chloride being laid down are tiny but incrementally can result in the formation of large crystals. That is also why crystallization can be a very slow process. If brine dries out too quickly only very small crystals form.

The role of order from chaos in the self-assembly of other molecules relevant to this story is described in the formation of vesicles from fatty acids and in the assembly of peptides on templates.

Self organisation

Obviously crystals are not living systems but they might have formed a template onto which polymers could have formed as Graham Cairns Smith suggests. The polymers may have consisted of polymerized amino acids (polypeptides) which the crystal template would have caused to line up in a consistent sequence to give a large amount of polymer product. Montmorillonite, a clay mineral formed by the weathering of volcanic ash, may have played a central role in the evolution of life. It can absorb organic compounds and this contributes to its ability to catalyse a variety of reactions critical to scenarios of life's origin. RNA binds efficiently to clays and montmorillonite can catalyse the formation of longer molecules, thus lending support to the hypothesis that life based on RNA preceded current DNA and protein based life.

Graham Cairns-Smith in his book, "Seven Clues to the Origin of Life" [22] postulates that:

i)   Genetic information is the only thing that can evolve through natural selection.

ii) DNA is far from the centre of biochemical pathways and so must be a late arrival.

iii) Earlier 'organisms', now missing, provided the scaffolding for modern genetic material

iv) Generations of organisms that precede modern ones may have had a succession of their own forms of genes that came and went.

v) Primitive replicative machinery could have been quite simple and easy to assemble in contrast with modern complex machinery

vi) Crystals are self-assembling structures and may have been low tech genetic materials

vii) Microcrystals in clays could have formed the first replicating templates and catalysts.

The following illustration of the structure of montmorillonite shows a regular aluminum silicate structure with the possibility of exchangeable cations.

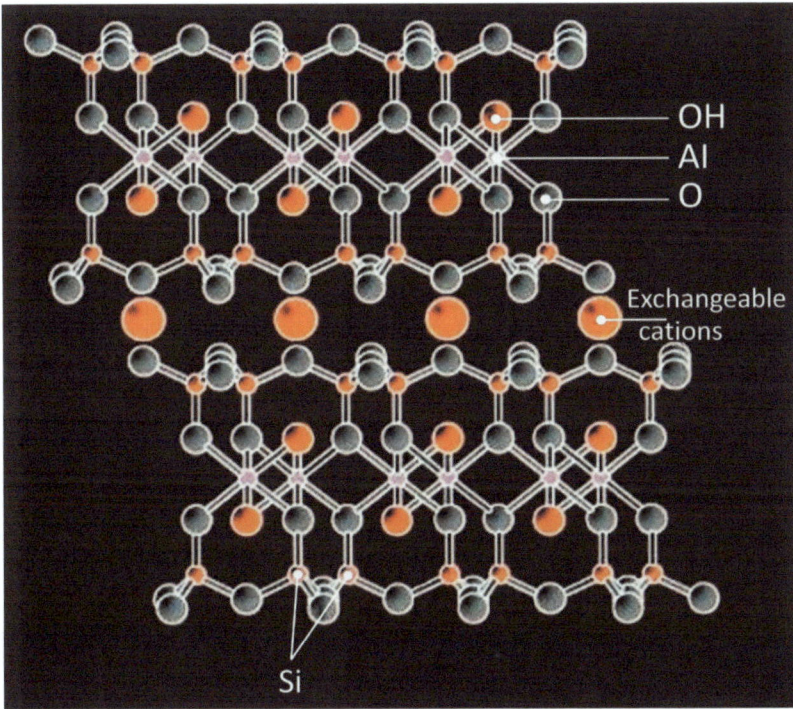

Figure 23 Structure of montmorillonite

The cations would supply points of electrostatic interaction with amino acids and similar compounds and this would allow the amino acids to line up in a consistent manner.

There are other molecular systems which tend toward self organisation and upon which the formation of life depends. Lipid membranes are such a system. Soap bubbles are an example of lipid micelles where the soap molecules have a hydrophobic (water repelling) end and a hydrophilic (water attracting) end . Hydrophobic means water hating or water repelling such as oily substances while the hydrophilic or water loving end is more prone to associate with water. When the soap molecules get together in a bubble they line up so that all of their hydrophobic ends associate with each other and all the hydrophilic ends associate with each other. This forms a continuous layer or can round up to form a spherical micelle, that is a membrane holding the bubble together and which forms spontaneously. This tendency of lipid molecules with both hydrophobic and hydrophilic ends to spontaneously form small bubbles called micelles is very likely the first step in a series of molecular events leading to cell formation. As an added benefit clay surfaces would confer the possibility of chirality to the peptides. This would make those peptides with enzyme-like activity very much more effective as catalysts.

Clay templates for polymer formation

As geologically relevant models of prebiotic environments, systems consisting of clay, water, and amino acids were subjected to cyclic variations in temperature and water content. Fluctuations of both variables produced longer oligopeptides in higher yields than were produced by temperature fluctuations alone. The results suggest that fluctuating environments provided a favourable geological setting in which the rate and extent of chemical evolution would have been determined by the number and frequency of cycles. One obvious way these cyclic fluctuations in the environment can come about is through tides. Kuhn, suggested in 1967 [24, 25] that if a pool containing some of these molecules is only flooded with more nutrients once per month at high tide then the intervening drying out period could have concentrated the nutrients and allowed them to form polymers or complexes.

Formation of vesicles in fire and ice

Membranes are important for metabolic processes. Separating the amino acid building blocks of an enzyme from the finished product would have imparted an enormous advantage to the process of enzyme formation. This is what is now described in biology as compartmentalisation.

> An early enzyme eg a replicase which essentially makes copies of molecules, probably would not have a built-in way of differentiating between a replicase or non-replicase sequence, and as a result, might make a copy of any substrate that happens to be close by. Without some means of separating the replicases from the non-replicases, the population of replicases is unlikely to grow and prosper. This issue can be resolved if the replicases are placed within a compartment, such as a vesicle, which can physically separate the replicases from other substrates.

Vesicles, themselves a product of self assembly, assist in overcoming chaos or entropy. Membranes allow reactants to remain in close proximity to each other rather than drifting off and becoming diluted. So if a reaction is going on as may have occurred where a catalytic peptide catalyses its own formation, the buildup of concentration of the amino acids entrapped in a vesicle would aid in the process by increasing the concentration of reactants.

In addition, a membrane may have played an important role in the early cell's ability to store energy in the form of a chemical gradient. In modern eukaryotic cells, the mitochondrion, often called the "cellular powerhouse" uses an internal chemical gradient across its membrane to create the energy-storing molecule ATP. The conversion is performed by a large enzyme called ATP synthase which rotates as a nano machine generating ATP molecules as it does so.

Bacteria and archaea

It is not worth re running the tape back though all the species exhaustively for this exercise (Dawkins has already done it in the Ancestor's Tale) but it should be enough to know that all species have arisen from previous species through an <u>unbroken</u> line of descent. We can go back through dinosaurs, amphibians, fish and worms to ultimately look at algae and other small multicelled creatures but these in turn have developed from single celled organisms.

Ultimately the single celled organisms that we know as bacteria would have evolved from a precursor organism which is no longer present. Such is the pace at which bacteria evolve which can be seen by the rate that they adapt to entirely new food sources or other challenges. Therefore it is hard to determine which path their evolution may have taken. However, there is a very strong clue in the existence of another completely separate class of single celled organisms known as the archaea.

Archaea

Now classed as a separate family from the bacteria and from higher organisms, archaea have a set of unique chemical components which set them apart from bacteria. They are found in highly unusual environments and for that they are also dubbed extremophiles. For example they can be found in huge numbers in soil, in hot springs, in highly acidic water and highly saline water. The features of these organisms that allow them to withstand these extreme conditions are not well understood but include, for example, the use of protein internal structures that resist decomposition by heat. A unique feature of archaea is the composition of their cell membranes which differ from bacterial cell walls in being composed of isoprenoid structures as opposed to the fatty acyl triglyceride structures of bacteria and eukaryote (Fig. 24). In the triglyceride molecules fluidity is conferred by adding double bonds in the fatty acid structure. This is called a level of unsaturation and the greater the level of unsaturation the greater the membrane fluidity. In archaea the fluidity of their membranes is conferred by virtue of the branched chain structure of the isoprenoids.(Fig. 25).

Archaea membranes

Membranes found in the archaea differ significantly in structure from those found in the other main families of life forms.

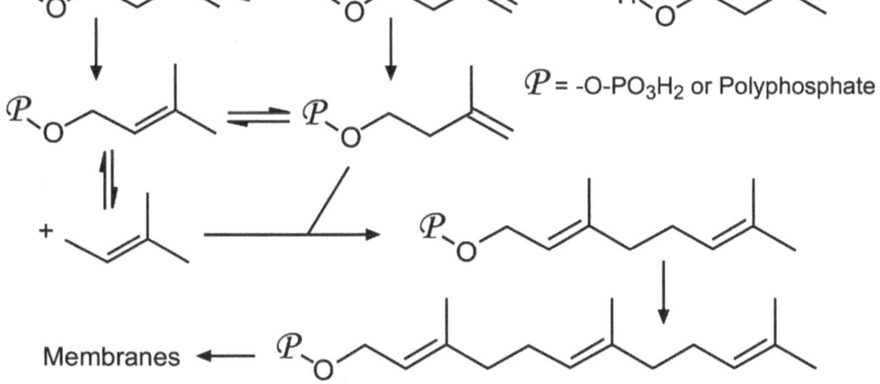

Figure 24, formation of isoprenoid structures in membranes of archaea.

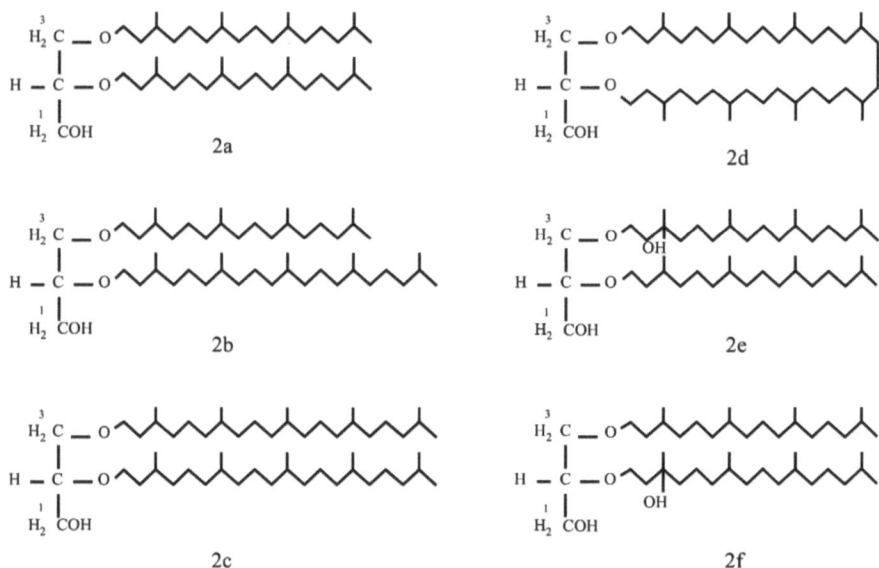

Figure 25, structures of ether lipids in the membranes of archaea

Formation of vesicles in fire and ice

Membrane vesicles allow reactants to remain in close proximity rather than drifting off and becoming diluted. So if a reaction is going on where an enzyme catalyses its own formation, the buildup of concentration entrapped in a vesicle would aid in the process of its own formation. Energetics for these reactions may have had its early origins in sulphur chemistry catalyzed by iron. Iron sulphur interactions can also produce the mineral FeS or iron sulphide which in its large crystalline form is known as iron pyrite or fool's gold. It has a brilliant gold lustre but can also form crystalline surfaces capable of binding some polymeric molecules.

Some scientists have proposed that hydrothermal vents may have been sites where prebiotically important molecules, including fatty acids, were formed. They propose a theoretical scenario in which fatty acids are formed along the face of a geyser. Research has shown that some minerals can catalyse the stepwise formation of hydrocarbon tails of fatty acids from hydrogen and carbon monoxide gas that may have been released from hydrothermal vents. Fatty acids of various lengths are eventually released into the surrounding water. The vents, if positioned next to snow or ice, as occurs in thermal hot springs during winter in Yellowstone National Park in Wyoming, USA would have provided an extra condition of alternate freezing and warming (which actually is a condition used to promote vesicle formation in the laboratory!!). Snow will not be cold enough to freeze the vesicles but mixed with a solute such as common salt, sodium chloride, (the freezing mixture used in a manual ice cream maker), would be cold enough to freeze vesicles. Other solutes such as calcium chloride are even more effective at producing freezing mixtures.

The fatty acids produced in this manner would only be found in low concentrations. Relatively high concentrations of fatty acids are required, however, to form higher order structures such as micelles and vesicles. Pools of water may have slowly accumulated fatty acids through cycles of shrinkage by evaporation and growth by the delivery of additional dilute fatty acid solution. It is also possible that droplets of fatty acids may have become aerosolized, such as being thrown into the air by a geyser, allowing the dry fatty acid particulate to travel long distances away from its original site of synthesis. Over time, small pools of water downwind may have accumulated high concentrations of fatty acids.

An experiment to show that fatty acids and similar molecules were available was to use the Murchison meteorite. A piece of the meteorite was extracted with organic solvents and the extract was then dried down and taken up in water (See figure 26). The resultant mixture yielded vesicles made of the material extracted from the meteorite, fatty acids and phospholipids. These materials allow the formation of vesicles which are essentially non-living cells. It is also possible that there were sources of fatty and oily deposits in large amounts of oily material from meteorites and even as a mineral in the Earth.

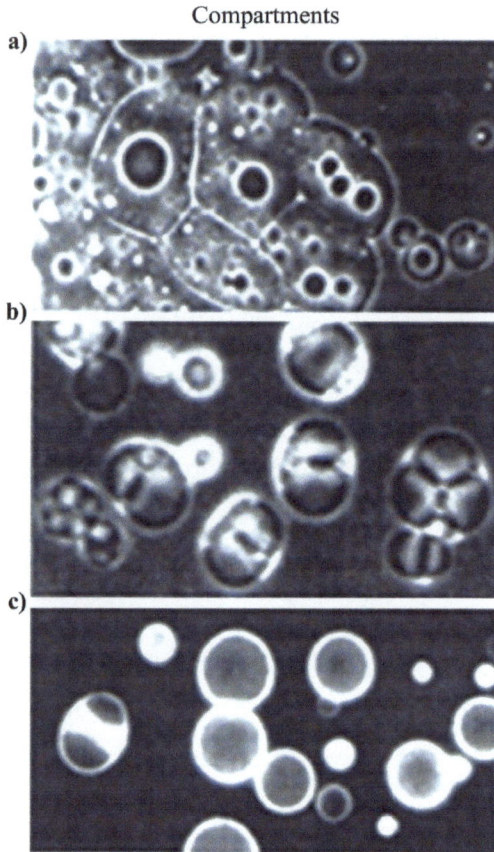

Figure 26, Vesicles from an extract of the Murchison meteorite

Vesicles have interesting properties in that they can fuse with each other or split apart. While doing so they may mix their contents such that new combinations of peptides and energy reactions could occur. This would make it possible to produce vesicles that contain several of the ingredients needed for making a functioning cell.

(a) Self-organization

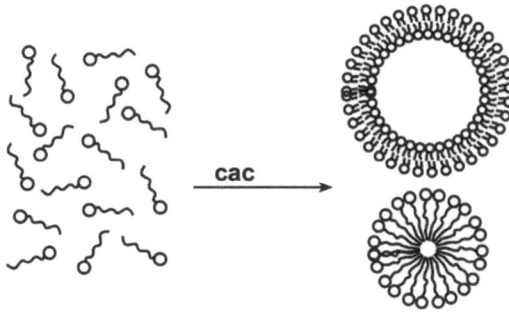

cac

(b) Compartmentation

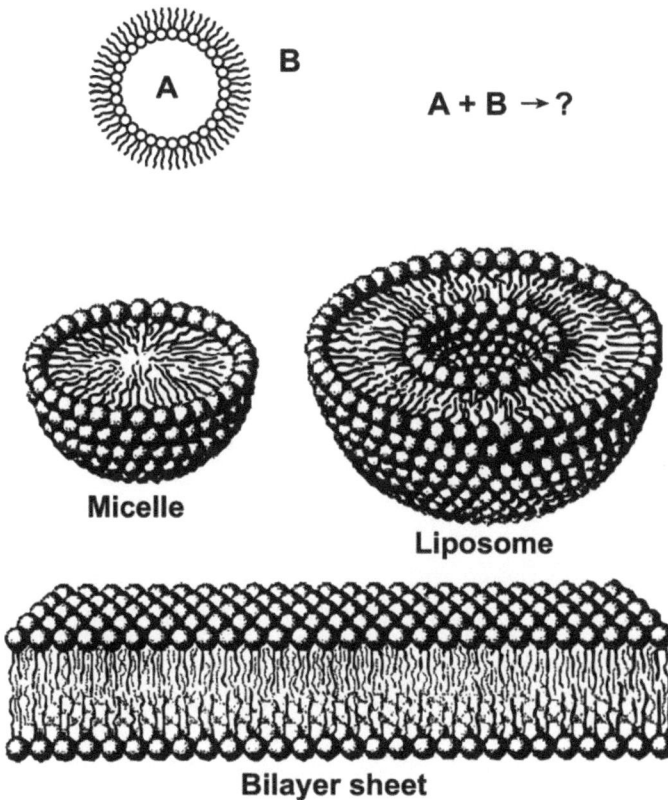

B

A

A + B → ?

Micelle

Liposome

Bilayer sheet

Figure 27. Self-assembly of vesicles

Conclusions: Order from chaos in the formation of vesicles allows the process of molecular self-assembly to get under way.

Chapter 3

PLATE TECTONICS THAT GENERATE THE FIRE

T he Earth has been undergoing massive shifting of continents over millennia. It has been through periods where there once had been a single continent called Pangea and then to break up into new smaller continents forming and reforming continents along the way. It took a long time for the theory of plate tectonics that drives this process to be accepted (Fig. 28).

> The "Father of Plate Tectonics", **Alfred Wegener** proposed "Continental Drift" in 1912, but was initially ridiculed although he was eventually vindicated.

The initial clue was the apparent fit of some edges of continents such as the east coast of Africa seems to mirror the west coast of south America. It was eventually shown that the movement of continents was driven by the upwelling of magma from deep below the Earth's crust coming up to form mid oceanic ridges of upwelling magma which push apart the continents. The upwelling of magma is a result of tidal forces from the sun's gravitational pull impacting the Earth by squeezing it in several directions. As the Earth rotates around the sun in an elliptical orbit it is undergoing changes in the gravitational pull from the sun. The pull gets weaker as the orbit takes the Earth further from the sun and stronger when it is closer. The moon also contributes to this effect and together they cause great amounts of heat to be generated in the layer below the Earth's crust as they squeeze the Earth like a ball of putty.

This magma layer is molten so that the crust, which contains the continents, floats above the magma layer. This causes the drifting of continents and causes such movements of the tectonic plates that they collide in places, the most well known subduction zone of which is the San Andreas fault along the north coast of California. The Pacific Plate undergoes subduction by sliding underneath the Californian plate and in doing so causes local magma heating. As a result, a row of volcanoes has sprung up along the west coast of USA. Mt St Helens is one of these which erupted in 1980. Volcanoes are active on all sides of the Pacific ocean making a region of instability along the coastlines and producing volcanoes. This active zone around the edge of the Pacific ocean is called the Ring of Fire

Plate tectonics

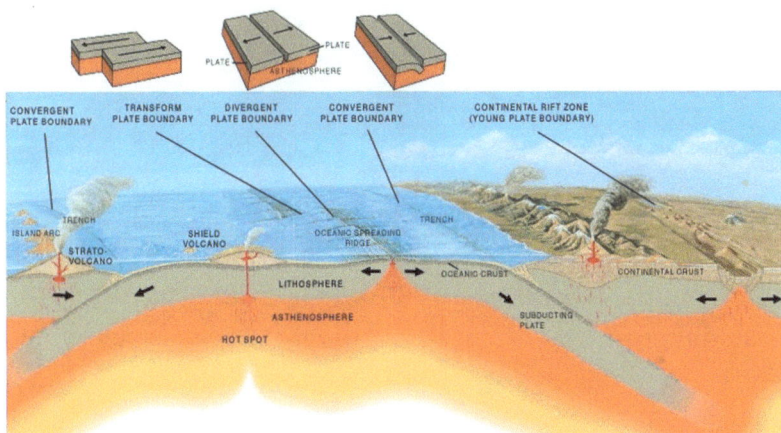

Figure 28 plate tectonics

The upwelling of magma in the mid oceanic ridges comes up as very hot molten rock meeting the cold of ocean waters which causes spectacular plumes of black smoky looking hot material jetting out of tube-like structures. These structures are called hydrothermal vents (Fig.29) or more colloquially black smokers pouring out minerals and chemicals into the ocean.

Hydrothermal vents in the deep ocean

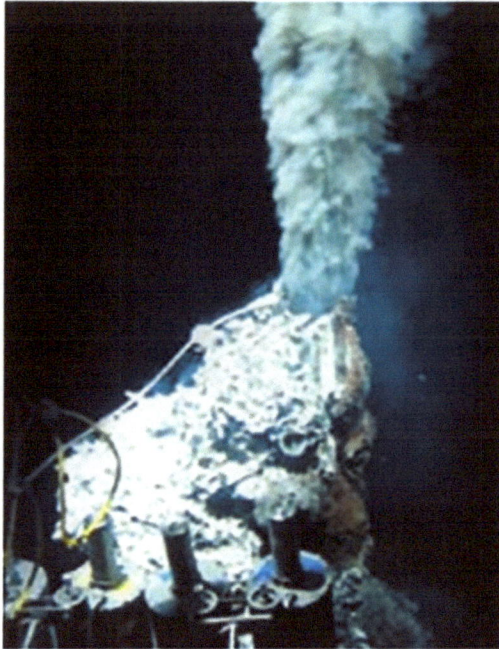

Figure 29 Black smoker vent resulting from hot magma welling up through a fissure in the bottom of the ocean.

The remarkable thing about these black smokers is the amount of life surrounding them, much of which gets its energy from heat and from sulphur chemistry rather than sunlight which doesn't penetrate water very far down. The smokers are home to unique strains of bacteria as well as beautiful red tube worms (Fig.30) that feed on the bacteria.

Figure 30, red tube worms inhabit the region near the black smokers.

Hoards of unique small crabs cover some of these areas (Fig. 31).

Dense mass of anomuran crab Kiwa around
deep-sea hydrothermal vent

Figure 31, tiny crabs swarming over a black smoker

The ultimate effect of the magma coming out is to push apart
continents In some places the mid oceanic ridges have come up

on land as in the case of Iceland where new land is being formed all the time. A new island of Surtsey (Fig. 32) was formed off the coast of Iceland in 1963. Similarly a new island is currently under construction to the north of the big Hawaii island but is yet to break through to the surface.

Figure 32 Formation of Surtsey island off the coast of Iceland.

Some island chains are the result of the tectonic plate moving over a hot spot which breaks through occasionally. An example is the string of islands of Hawaii where new islands are continually created. As the plate keeps moving a chain of islands results initially popping up as new volcanoes. Strings of islands off the coast of New Zealand is another example of this.

Yellowstone national park (in Wyoming USA) is a unique area of geothermal activity because it sits on a supervolcano which is magma spread over a large area underground but close to the surface because the Earth's crust is thin there. This results in a series of hot springs where the heat wells up from below turning water into steam. Spectacular geysers are everywhere in the park (Fig. 33) along with hot springs.

Figure 33, spectacular geyser activity in a hot spring

The hot springs are dangerous, some of them being close to the boiling point of water. Some have skeletons of animals that have fallen in. The whole area gets blanketed in snow during winter and the proximity to hot springs and geysers creates the environment of fire and ice giving the conditions which will be referred to several times in this book. There are many places in the world where hot springs occur and some of them also have ice and snow surrounding them.

The fire and ice conditions are right for several features of prebiotic reactions. They allow the formation of vesicles, for the binding of amino acids to a clay or mineral surface and making and remaking of material to be copied. They provide the heat conditions that can speed up reactions dramatically greatly enhancing the rate of evolution of chemical reactions. One observation often made is that life appeared too quickly for it to have happened by "chance". However, having reactions speed up because the ambient heat from hot springs could account for the rapidity of LUCA making an appearance. They also allow the breaking and remaking of base pairs in a DNA -like molecule, (see description of PCR reaction).

Deep vents vs hot springs

Many early life researchers propose black smokers as a likely place where life began. Deep vents or black smokers are compatible with extremophiles, sulphur based energetics. While they teem with life, they are too hot for stable polymers and clays are unlikely to be present.

Hot springs have the juxtaposition of hot and cold thermal cycling. This would assist formation of micelles. Clays are potentially present, accompanied by sulphur based chemistry all of which can generate the conditions for LUCA . Comparison of the two venues as likely places for first steps towards LUCA (Fig. 34).

Deep vents vs hot springs

- **Deep vents or black smokers**
- Compatible with extremophiles
- Sulphur based energetics
- Teeming with life
- Too hot for stable polymers
- Clays unlikely

- **Hot springs**
- Juxtaposition of hot and cold-> thermal cycling
- Formation of micelles
- Clays potentially present
- Extremophiles
- Sulphur based chemistry

Figure 34 Deep vents vs hot springs

Conclusion: Plate tectonics made possible the hot springs where it is proposed that life began.

Chapter 4

FORMATION OF PEPTIDES, NUCLEOTIDES AND RNA

W hat is a peptide? The joining of two amino acids by the formation of an amide bond between them and the elimination of the elements of water, oxygen and two hydrogen atoms.

The reaction for the formation of a peptide goes like this:

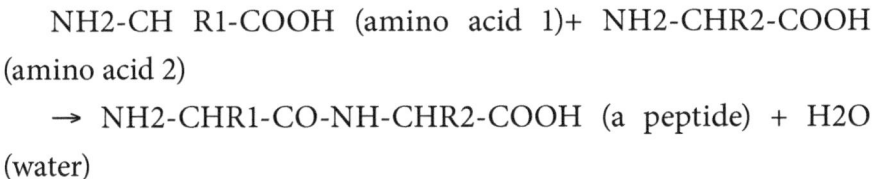

NH2-CH R1-COOH (amino acid 1)+ NH2-CHR2-COOH (amino acid 2)

→ NH2-CHR1-CO-NH-CHR2-COOH (a peptide) + H2O (water)

Where R1 and R2 represent different side groups, which can be aliphatic, hydrophobic, hydrophilic, charged negatively or positively or structural, allowing the amino acid to take on different chemical characteristics. Graphically presented as (Fig 35).

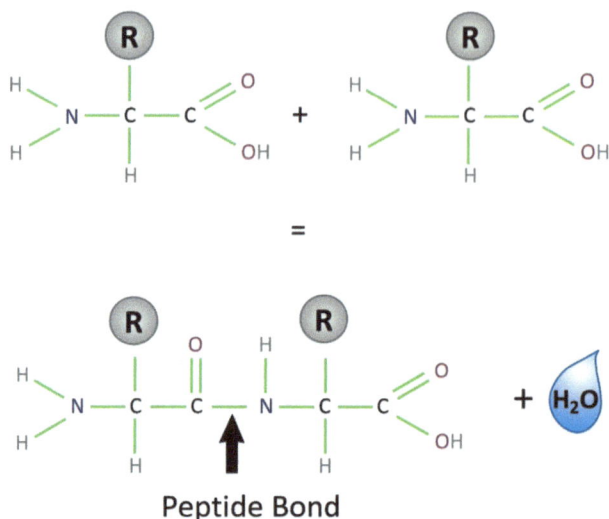

Peptide Bond

Figure 35, peptide bond formation

The equilibrium of the reaction required to form a peptide bond is unfavourable ie it pushes the reaction in the opposite direction to that of peptide formation. The equilibrium of the reaction favours peptide bond hydrolysis to form individual amino acids, not the peptide bond formation direction. However, the peptide bond once formed is very stable as it has the property of a 'delocalised' double bond (sharing electrons between three carbon atoms) allowing it to be rigid and stable. In water, solutions of peptides do not readily hydrolyse to form amino acids but this could be achieved using either acid or alkali conditions to promote hydrolysis. Therefore, how do peptides form spontaneously? The formation of a peptide involves the loss of the elements of water H and OH from the bond. One way to promote the loss of water is to use dehydrating conditions. These can be heat and drying, or dehydrating agents such as salts which absorb water.

Protein chemists use a chemical called a carbodiimide which acts to remove the elements of water from two amino acids to create a new peptide bond. In thermal springs the water can often be extremely acidic, as low as pH 0.2, with sulphuric acid present.

> The chemistry of hot springs is variable; they can range from water that is highly acidic (as low as pH 0.2) to very basic (pH 11). Walter K. Dodds, Matt R. Whiles, in Freshwater Ecology (Second Edition), 2010

Concentrated sulphuric acid is extremely dehydrating and could easily promote peptide formation by removing the elements of water from amino acids. If this is done inside a lipid vesicle with a template of some kind the peptides formed can be specific ones. The Cairns Smith suggestion is that the template is a clay which has regular silicate structures incorporating ionic sites which may have provided binding sites for particular amino acids thus allowing acidic conditions to promote particular peptide formation.

Peptides may form by simple approximation where one amino acid binds and sits next to another amino acid. The overlap of the two molecule ends where the atoms of water are present then allows for the acid catalysis or other dehydration process to remove the elements of water (OH from one reactant and H from the other reactant) by forming a bond between the two amino acids. Others have suggested pyrite or other mineral surfaces to catalyse polypeptide formation. eg Vincent L et al in Life, 2019, vol9, p80,[25]. Chemical Ecosystem Selection on mineral surfaces reveals long term dynamics consistent with the spontaneous emergence of mutual catalysis.) ref [26]

One study suggests a combination of six minerals was essential to provide an electrochemical disequilibrium deemed essential as an initiator of a CO_2 reducing metabolism [26].

Enzymes of the protein kind are made of long strings of peptides called polypeptides. Proteins can contain as few as 9 amino acids as in the case of some peptide hormones (eg oxytocin) or with its length of ~27,000 to ~33,000 amino acids (depending on the splice isoform), **titin** is the **largest** known protein. These polypeptides do not remain as a linear arrangement like beads on a string but form local clumps as they are made and eventually fold into a complex three dimensional structure which, for enzymes, has catalytic surfaces within its structure. I liken this clumping to the effect of winding up a rubber band in a model plane. As it is wound the initial spiral shape begins to clump up. As you keep winding, if it doesn't break first, the clumps begin to form bigger clumps. These peptide clumps form catalytic surfaces in enzymes which act to catalyse one specific reaction within the organism. So there are thousands of these enzymes in our bodies each performing mostly a single specific tasks to operate our metabolism. Each one has its own unique task to perform to convert a metabolite into another metabolite as part of a metabolic pathway. It can be shown through looking at the structure of conserved components, domains, within these structures that enzymes have evolved from other enzymes. The gene structures are composed of identifiable domains which may provide specific functions such as binding sites for a particular effector molecule or for attachment to a surface or delivery to a particular subcellular compartment. New enzymes may evolve by incorporating a domain from a precursor enzyme and adding new domains to create a new enzyme. In the beginning it would not have been as complex. Other modern biochemical roles for proteins include muscle proteins, hair keratins, oxygen transporters haemoglobin and myoglobin, collagen in connective tissue and many others.

Some theorists believe that proteins did not come first but that the information contained within DNA and RNA had to come first because that is what happens now. But there is a strong argument that proteins probably did come first. A primary reason is that the RNA molecule is complex and is unlikely to have formed spontaneously. RNA is more likely to have needed some kind of catalytic assistance to form such as that provided by protein based enzymes. In fact, once RNA enzymes became possible, the primary entity forming new catalytic surfaces may have changed several times from a peptide based catalyst to a nucleotide or RNA based system, each supporting the other by an association which protected against UV damage and other external forces.

If this occurred within an acidic environment, it is possible as vesicles were washed around, that the external environment in which the vesicles found themselves was water. If the internal solution was still acidic this would have allowed a proton gradient to be set up across the vesicle membrane and this can drive chemical reactions. Other possible energy sources could be the hydrolysis of polyphosphates and /or the energetics provided by iron/sulphur complexes. The proton gradient could be a precursor to the later generation of ATP from proton gradients across membranes that occurs now in mitochondria .

An extensive review by Tessera [17] explores many models of prebiotic chemistry including metabolism-first models (eg. peptides) and replicator-first (eg. RNA) scenarios. While none of the theories satisfy all of the criteria and leads Tessera to the conclusion that there was no such thing as pre-Darwinian evolution, it does not preclude some kind of chemical progression which must have occurred albeit not as a classical evolutionary process.

Another proposal [27] suggests the origin of life occurred in alkaline hydrothermal vents (rather than acidic). To quote the authors "alkaline vents are proposed to have acted as electrochemical flow reactors, in which alkaline fluids saturated in H2 mixed with relatively mild ocean waters rich in CO2, through a labyrinth of interconnected micropores with thin organic walls containing Fe(Ni)S minerals. The difference in pH across these thin barriers produced natural proton gradients with equivalent magnitude and polarity to the proton-motive force required for carbon fixation in extant bacteria and archaea."

An area that must also be explored is the origin of chirality. This is the observation that many biological molecules have handedness, eg amino acids come in an L form and a D form which are not superimposable just as your left hand and your right hand cannot be copies of one another but are mirror images. Most proteins consist of exclusively L amino acids although a few small peptide hormones do contain D amino acids. One way that handedness can easily occur is if there is a template, such as, but not necessarily clay, onto which amino acids are laid down. The surface reduces the number of degrees of freedom thus making it more likely that all the amino acids laid down would have the same handedness of orientation. This is the reason I prefer the clay template theory even though some theorists argue against it.

Conclusions

An environment of fire and ice in a hot spring surrounded by snow and ice could generate vesicles which also have chemical or physical gradients creating an energy source. The low or high pH, highly dehydrating environment drives peptide synthesis in the vesicles and, in the presence of clay or iron pyrite templates, polypeptides can result which are defined in their sequence structure. Alternative chemical scenarios are possible and explored in [17].

POLYPEPTIDES WITH ACTIVITY
Self reproduction by limited peptides

Stuart Kauffman (The Origins of Order, 1986) [28] showed that:

peptides that are randomly allowed to form from a mixture of amino acids **will generate some peptides with catalytic activity.** Initially the only ones that increase in amount in this mixture are peptides that catalyse their own formation. However, in our primitive vesicle this is not a perfect copying environment and 'mistakes' will occur. Nor is it a pure solution of only amino acids so other molecular interactions are almost certain to occur. Particularly hydrophobic associations which offer mutual protection against decomposition by agents such as UV irradiation. Subsequent mistakes/mutations/changes can be readily envisaged in these self-catalysing peptides to allow them to catalyse other reactions such as bond breaking and bond making to make other peptides, or bonds between other components like sugars, that catalyse new reactions and make new peptides and other compounds. Associations with other molecules such as purines and pyrimidines, metal ions and mineral salts would have influenced the products formed in this environment. While complex associations may have occurred, the process of enriching a population of particular peptides will have had the effect of partially purifying that peptide in the mixture.

To quote Kauffman in reference to the randomly forming peptides

"This leads to the emergence of self replicating systems as a self-organizing collective property of critically complex protein systems in a prebiotic evolution".

So peptides could have developed with specific self-replicating activities. Indeed this is the only way that we can envisage single peptides increasing in amount from within a mixture of random peptides. The 'enzyme' activity displayed by the peptide would involve peptide formation and as 'mutations' occur in this highly impure and complex mixture the catalytic activity could transform into catalysing the formation of other bonds. The smallest known enzyme of the current biochemistry is a peroxidase of 60 amino acids in length. Primitive enzymes may have been shorter. The peptides in development here are the ones referred to as ancestral peptides which have specific roles such as binding nucleotides.

Peptides with activity consist of associations with other components such as nucleosides eg adenine, and sugars, which, by virtue of their aromatic structures provide protection against UV damage and free radical damage. Vesicles containing these peptides fuse with other vesicles containing perhaps different peptides or other complementary compounds such as polyphosphate. Polyphosphate and pyrophosphate (diphosphate) both release energy when hydrolysed and could help to provide energy for the vesicles. Other iron sulphur reactions could be part of the mix [21] and eventual efficiency improvements by covalent linkages leads to proto-enzymes from both peptide and nucleoside origins. While peptide based enzymes would initially be limited there could be a supporting reaction driving the formation of particular bonds such as sugar-purine bonding, particularly ribose, supported by polyphosphate energetics. Eventually the phosphate becomes incorporated because this is more efficient and hence you have development of **the first nucleotide**.

Conclusion: A crucial step in the formation of peptides is that some with catalytic activity are produced. Catalytic peptides can plausibly lead to the formation of nucleotides.

Nucleotides

Structually a nucleotide is a triphosphate linked to a sugar (ribose) which is in turn linked to one of four possible purine or pyrimidine bases (Fig.35),

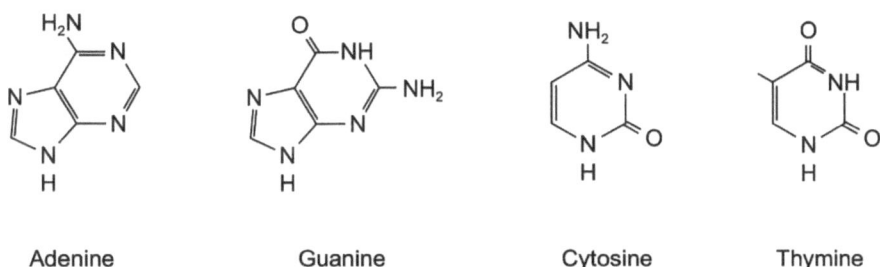

Adenine   Guanine   Cytosine   Thymine

Figure 35, purine and pyrimidine base structures

ATP, below (fig.36), is the archetypical nucleotide where adenine is linked to ribose which in turn is linked to a triphosphate.

Adenosine triphosphate (ATP)

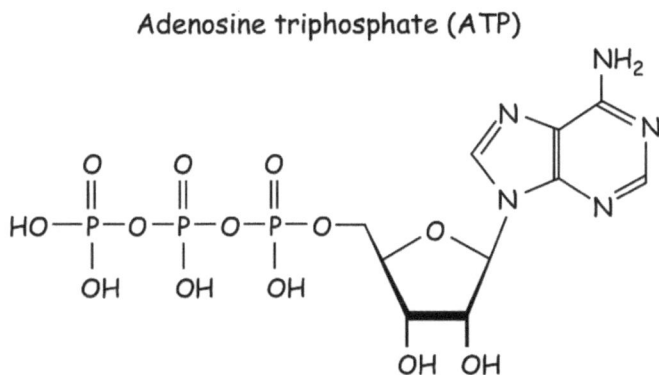

Figure 36, structure of ATP

Note that in DNA the nucleotide bases can be Adenine , as shown, or Guanosine, Thymine or Cytosine. These are linked to the sugar ribose which is also linked to a triphosphate. The bond between the two end phosphate molecules is high energy and when hydrolysed releases energy that powers most of the energy requiring reactions in the body. There is no way this molecule could have arisen spontaneously.

These single nucleotides can then polymerize to form RNA, although one of the bases in RNA is uracil (Fig. 37) in place of thymine, and so the RNA catalysis of a now larger array of reactions can occur bringing in the RNA world scenario.

Uracil

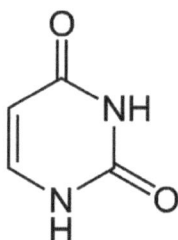

Figure 37, structure of uracil

It is possible that the first such RNA enzyme or ribozyme was one which made RNA more efficiently than the random pathway I just described. A more efficient pathway would be readily adopted and would take over such that traces of the original pathway are totally obliterated. In modern biochemistry the process is again taken over by polypeptide based enzymes except in the case of the peptidyl transferase activity of ribosomes.

The RNA molecule

RNA is a somewhat complex molecule made up of a polymer of individual ribose-sugar-base components linked together through phosphate moieties. It is certainly highly unlikely, indeed not remotely possible, for it to have arisen spontaneously from its components.

The Emergence of RNA from the Heterogeneous Products of Prebiotic Nucleotide Synthesis

Seohyun Chris Kim [12], Derek K O'Flaherty [13], Constantin Giurgiu [12], Lijun Zhou [13], Jack W Szostak [123]: 10.1021/jacs.0c12955

Shapiro has shown that: "Many accounts of the origin of life assume that the spontaneous synthesis of a self-replicating nucleic acid could take place readily. Serious chemical obstacles exist, however, which make such an event extremely improbable. Prebiotic syntheses of adenine from HCN, of D,L-ribose from adenosine, and of adenosine from adenine and D-ribose have in fact been demonstrated. However these procedures use pure starting materials, afford poor yields, and are run under conditions which are not compatible with one another. Any nucleic acid components which were formed on the primitive Earth would tend to hydrolyze by a number of pathways. Their polymerization would be inhibited by the presence of vast numbers of related substances which would react preferentially with them. It appears likely that nucleic acids were not formed by prebiotic routes, but are later products of evolution." Orig Life, 1984;14(1-4):565-70. [29]

A lot of cerebral energy has been expended trying to work out how this RNA might have arisen in an early Earth. However, it makes more sense, to me, to have peptides with activity generated first and then RNA formation catalyzed by enzymes as it is today.

RNA differs from DNA in that it can form 3 dimensional shapes whereas DNA is generally confined to forming the double helix shape found within chromosomes. The only difference in the structures is the hydrogen on the ribose sugar in DNA replacing the OH in

RNA. Hence the name 'deoxy' ribose nucleic acid giving us the 'D' of DNA. This difference allows the two molecules to behave differently in folding. While RNA can form 3D structures allowing catalytic surfaces to form, it may also form based paired regions allowing pairing with single stranded DNA and with regions within itself or with substrates as found in ribozyme activities. This makes RNA a very versatile molecule, having both information storage in its sequence and catalytic surfaces in its folding. The differences in RNA and DNA are demonstrated in Fig.38 and RNA in more detail in Fig. 39.

DNA vs. RNA

a.

Double-stranded   Generally single-stranded

b.

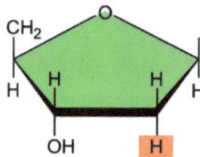 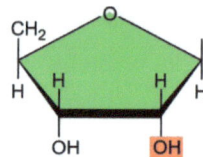

Deoxyribose as the sugar   Ribose as the sugar

c.

Bases used: Bases used:

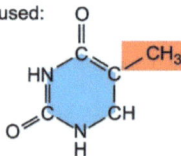 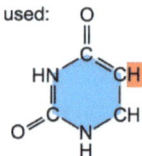

Thymine (T)   Uracil   (U)
Cytosine (C)   Cytosine (C)
Adenine (A)   Adenine (A)
Guanine (G)   Guanine (G)

d.

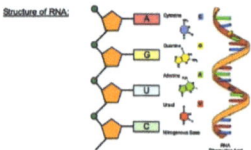

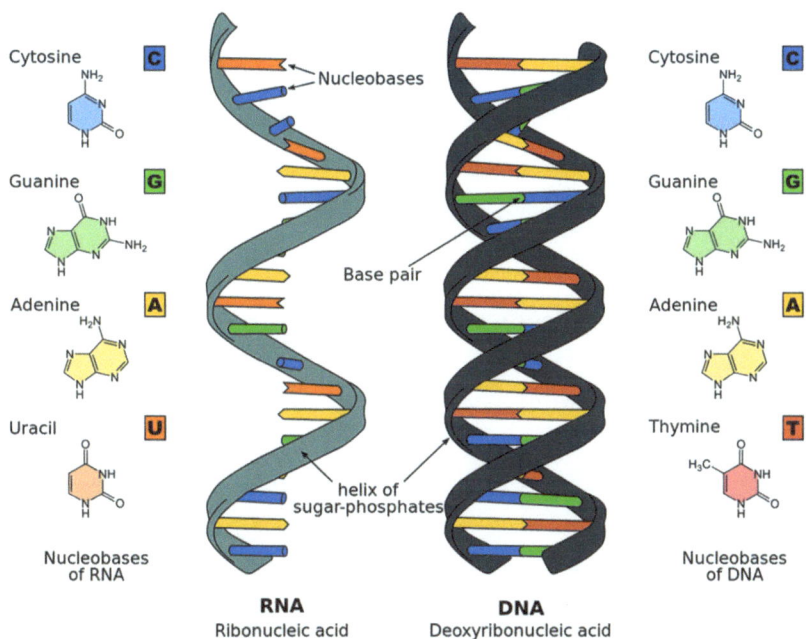

Cytosine **C**

Guanine **G**

Adenine **A**

Uracil **U**

Nucleobases
of RNA

Nucleobases

Base pair

helix of
sugar-phosphates

RNA
Ribonucleic acid

DNA
Deoxyribonucleic acid

Cytosine **C**

Guanine **G**

Adenine **A**

Thymine **T**

Nucleobases
of DNA

Figure 38, comparison of DNA and RNA structures

Structure of RNA

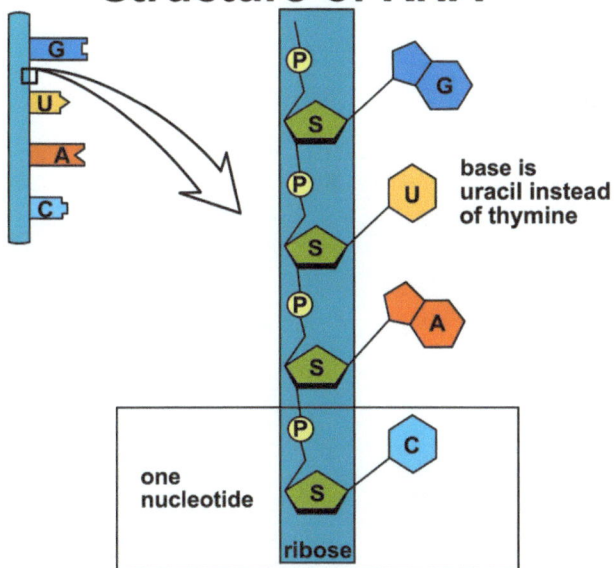

G

U

A

C

P

S

G

P

S

U

base is
uracil instead
of thymine

P

S

A

one
nucleotide

P

S

C

ribose

Figure 39, structure of RNA

The 3D shapes of RNA led some to investigate possible catalytic activities and this led to Sydney Altman and Tom Czech receiving the Nobel prize for their observations concerning ribozyme activity. An observation by RH Symons et al. of hammerhead ribozymes causing self cleavage. (see ref below) should have also shared the prize.

In modern biochemistry there are several forms of RNA within biological systems each of which has a different shape and function.

Transfer RNA (tRNA): (Fig.40) is a small form (70-90 nucleotides) of RNA which carries a single amino acid to the site of peptide bond formation in the Ribosome complex. It also contains a triplet code recognition sequence, known as the anticodon, allowing it to recognise the next codon triplet position on the DNA sense strand which directs the amino acid it carries to the next position in the growing peptide (Fig. 41).

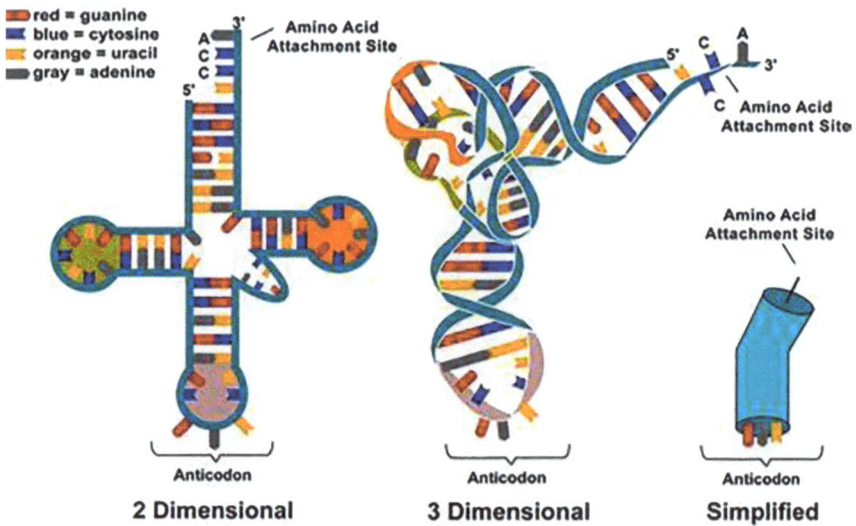

Figure 40, structure of transfer RNA

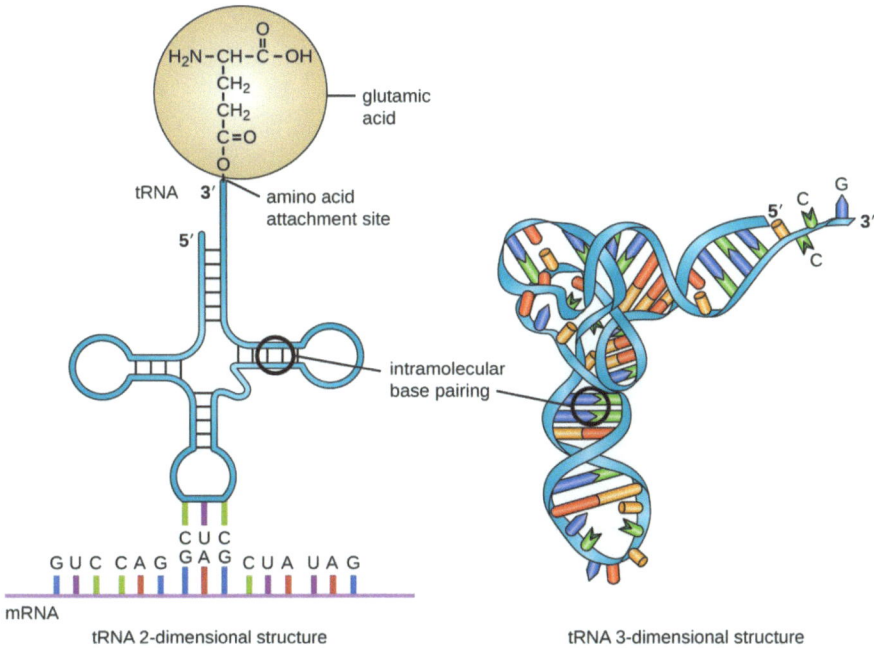

Figure 41, functional role of tRNA

Messenger RNA (mRNA): a longer form of RNA (depending on the gene length) made from a region of DNA containing the gene and which has the entire message to make the protein which that gene codes for plus some intervening sequences. Once the genomic messenger RNA has been copied and made from the DNA it must then be processed as it typically contains non translatable regions called introns. These must be excised out and the regions to be translated, called exons are ligated (joined) together to make the final Messenger RNA which is then presented to the ribosome for translation into protein.

Ribosomal RNA (rRNA): The ribosome is a large conglomeration of peptides and RNA present in every cell. The ribosome makes proteins by translating the message from DNA mediated by messenger RNA.

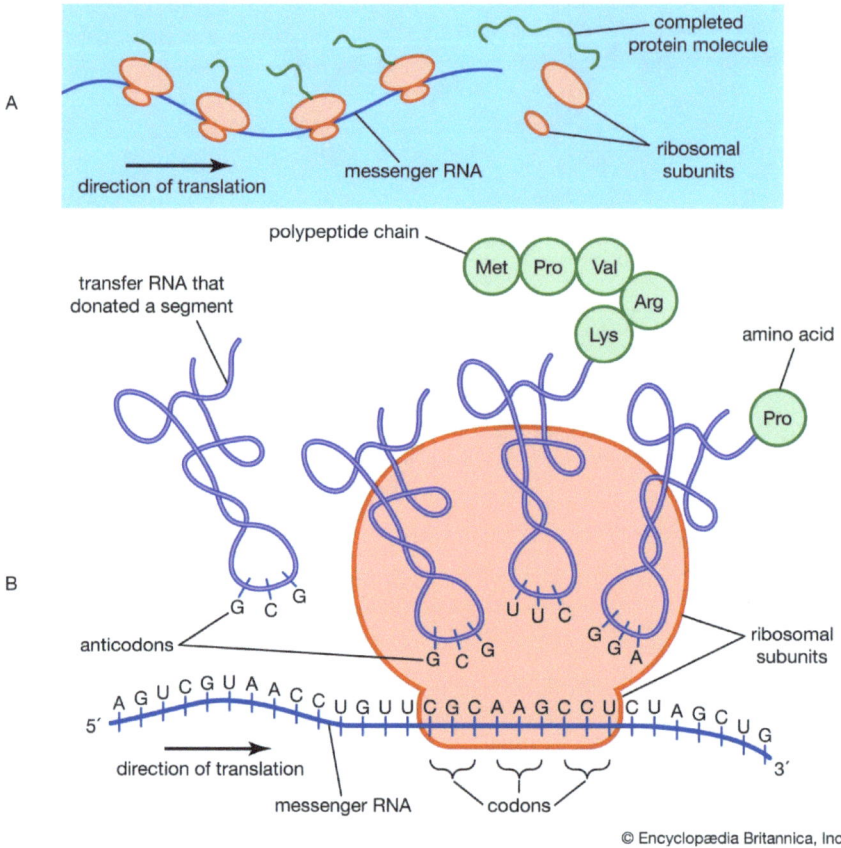

Figure 42, showing the action of the ribosome in translating messenger RNA into peptides with mediating tRNA molecules supplying the amino acids and reading the code.

The amino acids are carried to the ribosome by another RNA molecule tRNA (see fig. 42 for the ribosomal activity). There is a different tRNA for each triplet codon that codes for an amino acid. The RNA component of ribosomes is contained within active site sections of a ribosome in tight association with ribosomal peptides. It has been shown that the peptidyl transferase activity by which peptides are formed within ribosomes is catalyzed entirely by the RNA component (Fig. 43) of a ribosome. So the ribosome is a functioning ribozyme.

Figure 43, Sequence of ribosomal RNA from a bacterium, E coli.

Gene splicing RNA: RNA enzymes, ribozymes, that perform functions such as RNA splicing and editing. Some of these are encoded by intron DNA, ie the region between coding regions (exons) within a gene sequence. This region is cut out by processing enzymes, also ribozymes, when the messenger RNA molecule is made ready for translation into protein by ribosomes. This Ribozyme activity is the basis for a tool called gene shears which can cut DNA at specific sites.

Ribonuclease RNA: The enzyme ribonuclease P has been shown (Gardiner et al. [30]) to contain an RNA component within a protein complex and by stripping away all of the protein Altman's group showed that all of the catalytic activity is performed by the RNA component.

Ribozymes: Enzymes made entirely of RNA (eg. Fig.44). These enzymes have important functions in modifying DNA or RNA and are highly specific to a particular section of DNA or RNA by virtue of an internal guide sequence which base pairs with the section of the substrate to provide a precise site for its action of either cutting or splicing. There are many of these some of which are encoded by regions of DNA between protein coding regions called introns.

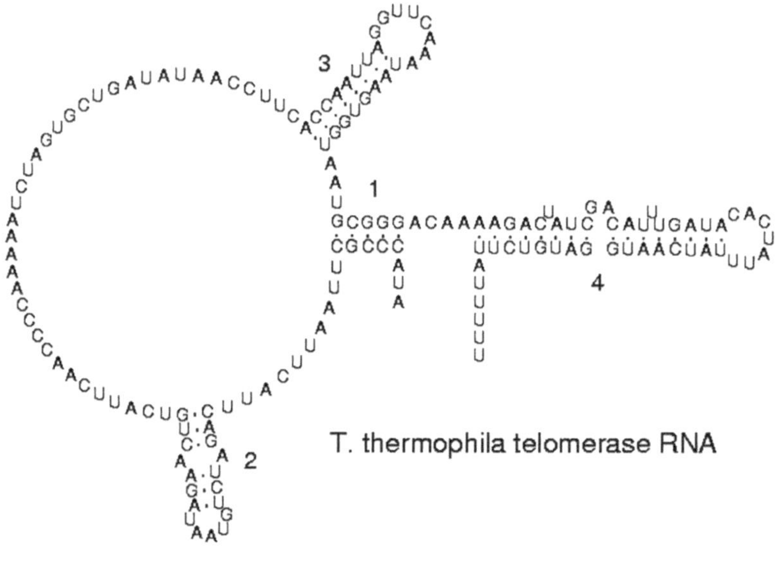

Figure 44, an RNA enzyme (ribozyme) which trims the ends of telomeres

Conclusion: Complex structures of nucleotides and ultimately of RNA are likely to have been made by enzyme activities, initially protein based and then probably RNA based.

RIBOZYMES AND THE RNA WORLD

Within the co-mixtures of peptides, purine or pyrimidine bases, sugars and pyrophosphate, there arise peptides with self-copying activity. This may be template guided on a clay or mineral surface. The range of bond formations increases with eventual formation of a ribose (or similar sugar)-adenine (or guanine or cytosine or uracil, all known as bases) linkage which again may be template guided. This ribose-base entity is called a *nucleoside* and it is generated by a peptide enzyme which has developed bond coupling between hydroxyl groups on different molecules. If polyphosphates are present there may also be a catalysed linkage with a triphosphate attached to the ribose to form ATP. This is now called a *nucleotide*. It may have incorporated one of the other purines or pyrimidines or perhaps it was a simpler base such as imidazole. The next step would be to couple each nucleotide to another nucleotide with the elimination of pyrophosphate. This then gives us something like RNA but I stress that it had to be in association with protective peptides., The RNA would be prone to hydrolysis as Shapiro points out above so I propose that peptide association would make it more stable. Formation of ribozymes with peptide support leads to ribozyme- based metabolic processes (the RNA world) . This RNA world has a limited set of catalytic functions which can produce a metabolism of a very simple kind. Eventually the association with peptides leads to a genetic code (See below for antisense explanation). Through many iterations a two base code with a third position occupied with perhaps an amino acid, or something else, becomes a three base code.

The RNA world

Enzymes were once thought to be made exclusively of proteins. However, Tom Czech, Bob Symons, Paul Berg and Sydney Altman working on different systems each made the discovery that certain RNA molecules have catalytic activity. RNA is different from DNA in having one extra oxygen on the ribose ring. This allows the RNA molecule to fold in three dimensions, unlike DNA which forms a rigid helix by virtue of base pairing. RNA however, can fold in many ways which allows it then to form catalytic surfaces just like a protein does. Sections of the folded RNA may also exhibit base pairing. It is this combination of properties that allows the formation of catalytic RNA's. Czech discovered that the Tetrahymena organisms had RNA activity which catalyses the cleavage of specific RNA molecules. The way this is done is to exploit the most obvious property of RNA that it will form base pairs. This is used to match up a special sequence, known as an internal guide sequence, on the catalytic RNA paired with the sequence to be cleaved. The RNA with catalytic activity is given the name ribozyme because it is an enzyme made of ribonucleic acids. Once the substrate RNA forms a base pairing with the internal guide sequence the ribozyme then initiates the cleavage through the use of a special binding site for a co-factor guanine molecule which aids in the catalytic process to cleave the RNA substrate. In later work it was shown that this activity could be modified to catalyse the formation of RNA-amino acid bonds such as those that would be encountered in transfer RNA molecules. It has also been shown that the peptidyl transferase activity that forms peptide bonds within ribosomes is catalysed entirely by ribosomal RNA.

Symons and co-workers discovered a self-cleaving RNA, named hammerhead RNA because of its shape [31]. Solutions of this molecule would appear to fall into smaller pieces when isolated leading to the conclusion that the molecule had self-catalytic activity. Variations of these molecules are now being used in genetic engineering of DNA and RNA molecules to make precision cuts and splices to molecules, the so called 'gene shears'. It is now thought possible to repair gene defects and errors in the tissues of patients with genetic diseases using an internal guide sequence approach. Kmiec (Ref) has done this using a simple RNA molecule carrying the sequence to be corrected which is added to the defective cells and uses the cells own repair enzymes to correct the defect using the introduced template.

The revolution of gene modification by a technique called CRISPR (Clustered Regularly Interspaced Short Palindromic Repeat) is based on these targeted DNA cutting and splicing approaches.

Altman's work centred on the internal workings of the enzyme ribonuclease P which digests RNA down to its constituent bases. This enzyme has a component of RNA within its structure. Altman was able to show that by removing the protein component of the enzyme, under proper conditions the RNA was responsible for all of the catalysis. The protein was therefore only present to stabilise the RNA and perhaps to bind to the substrate RNA. This may reflect the original relationship of nucleic acids and peptides with the RNA having the function and the peptide doing the protection.

These activities are only a sampling of the activities now found in RNA molecules and it is possible that an RNA catalysed metabolic system could have occurred in RNA based organisms. Fig below shows a list of the RNA catalysis types that have been discovered thus far. These findings and other led Walter Gilbert to propose a probable 'RNA world' era when only RNA-based organisms existed. This would have had to occur in the period before the appearance of bacteria. The RNA world would have ended abruptly when a new development occurred in the type of catalysis using proteins was introduced to the population of RNA organisms. Once this breakthrough occurred it would have taken over and obliterated the RNA world almost without trace. We will pull these threads together later into a theory of the genetic code.

However, in an excellent study looking at the intimate details of the RNA world hypothesis, Bowman et al [32] have questioned the basic premises of the RNA world hypothesis and called for a major update of the hypothesis. It is now thought that rather than naked RNA the era was one of an association between RNA and peptides. See below.

Production of Ribozymes

The modern remnants (molecular paleontology) of the RNA world are proposed by Visser and Kellogg to be the nucleotide coenzymes which have innate reactivity. Molecules we know today as NAD (nicotinamide adenine dinucleotide), ATP (adenosine triphosphate) Coenzyme A, and including the ribosomal RNA. Ribozymes show that catalysis by RNA is not only possible but it remains a common occurrence in modern biology (Fig. 45). However, the RNA world may never have existed in isolation, in fact it is my, and others, contention that it must have existed in association with protective amino acids and peptides and very possibly other compounds associated with this complex which helped stabilize and protect it from degradation.

Bond Formed	Leaving Group	Activity of Ribozyme
$-O-PO_3-$	5'-RNA	Phosphodiester cleavage
$HO-PO_3-$		Cyclic phosphate hydrolysis
$-O-PO_3-$	PP,	RNA ligation
$-O-PO_3-$	PP,	Limited polymerization
$-O-PO_3-$	AMP	RNA ligation
$-O-PO_3-$	ADP	RNA phosphorylation
$-O-PO_3-$	Imidizole	Tetraphosphate cap formation
$-O-PO_3-$	Rpp	Phosphate anhydride transfer/hydrolysis
$-O-PO_3-$	PP,	RNA branch formation
$-O-CO-$	AMP	RNA aminoacylation
$-O-CO-$	3'-RNA	Acyl transfer
$-O-CO-$	AMP	Acyl transfer
$-HN-CO-$	3'-RNA	Amide bond formation
$-HN-CO-$	AMP	Peptide bond formation
$-N-CH_2-$	I	RNA alkylation
$-S-CH_2-$	Br	Thioalkylation
$-HC-CH$		Diels-Alder addition (anthracene-maleimide)
$-N-CH$	PP,	Glycosidic bond formation

Figure 45, Some of the activities found in ribozymes

THE RNA-peptide world

Very good papers on the evidence for pre-Darwinian RNA -peptide associations include those by Kunnev and Gospodinov [33,34] where the authors explore the early development of the role of peptides in association with RNA to facilitate the activity of the RNA enzyme. Alva and Lupas [35] explore the early development of ancestral peptides which evolved as cofactors of RNA based replication and catalysis. The peptides involved are sometimes described as ancestral because they then turn up as domains in later evolved proteins. The ancestral domains probably had the role of specific association with some types of nucleotides or even of fully formed RNA molecules. Carter proposes that a RNA peptide -RNA partnership arose in order to interact with ATP [36].

Primitive selection of the fittest emerging through functional synergy in nucleopeptide networks Anil Kumar Bandela[1], Nathaniel Wagner[1], Hava Sadihov[1], Sara Morales-Reina[2], Agata Chotera-Ouda[1], Kingshuk Basu[1], Rivka Cohen-Luria[1], Andrés de la Escosura[3,4], Gonen Ashkenasy[5] 10.1073/pnas.2015285118

Many fundamental cellular and viral functions, including replication and translation, involve complex ensembles hosting synergistic activity between nucleic acids and proteins/peptides. There is ample evidence indicating that the chemical precursors of both nucleic acids and peptides could be efficiently formed in the prebiotic environment. Yet, studies on nonenzymatic replication, a central mechanism driving early chemical evolution, have focused largely on the activity of each class of these molecules separately. We show here that short nucleopeptide chimeras can replicate through autocatalytic and cross-catalytic processes, governed synergistically by the hybridization of the nucleobase motifs and the assembly propensity of the peptide segments. Unequal assembly-dependent replication induces clear selectivity toward the formation of a certain species within small networks of complementary nucleopeptides. The selectivity pattern may be influenced and indeed maximized to the point of almost extinction of the weakest replicator when the system is studied far from equilibrium and manipulated through changes in the physical (flow) and chemical (template and inhibition) conditions. We postulate that similar processes may have led to the emergence of the first functional nucleic-acid-peptide assemblies prior to the origin of life. Furthermore, spontaneous formation of related replicating complexes could potentially mark the initiation point for information transfer and rapid progression in complexity within primitive environments, which would have facilitated the development of a variety of functions found in extant biological assemblies.

Conclusion: RNA develops from nucleotide polymerization and gives rise to Ribozymes and the RNA world but in close association with specific peptides. The ancestral peptides develop into well known domains of modern proteins. The association of RNA with specific peptides gives rise to the RNA-peptide world.

Chapter 5

DEVELOPMENT OF THE GENETIC CODE

Collective evolution and the genetic code

Kalin Vetsigian, Carl Woese, and Nigel Goldenfeld

PNAS July 11, 2006 103 (28) 10696-10701; https://doi.org/10.1073/pnas.0603780103

Carl Woese and co-workers developed a stereochemical theory of the origin of the genetic code [37,38]. They postulated that individual amino acids have 'affinity' for individual nucleosides or pairs of nucleosides. The affinity is found of specific associations between nucleosides and amino acids mixtures and a crude code develops as dinucleotides have affinity for specific amino acids or short peptides. The affinity is based on shape complementarity and on hydrophobic interaction. A code develops out of this affinity relationship. The first 'code' was probably a 2 base code producing 10 amino acids. With no intermediate molecules to direct the process it would have been cumbersome and very inefficient.

Within the co-mixtures of peptides, nucleosides, sugars and pyrophosphate, there are peptides with self-copying activity. The range of bond formations increases with eventual formation of a compound something like RNA but in association with protective peptides.

This reverses the above peptide protection speculated above. Formation of ribozymes with peptide support leads to ribozyme-based metabolic processes (the RNA-peptide world) and the association with peptides leads to a genetic code. Through many iterations, or from a strong association of amino acids or dipeptides with nucleotide bases, a two base code (Fig. 46) , which may have coded for the ten ordinary amino acids, becomes a three base code. Speculative peptides from the two base code are shown (Fig. 47). As Crick pointed out it must always have been a three base code but perhaps the third position was not always occupied by a nucleotide base. See Fig. 48.

The Two Base Genetic code

1st position	U	C	A	G
	Second	*Position*		
	U	**C**	**A**	**G**
	Hydrophobic	**Hydrophilic**	**functional**	**structural**
U	phe	ser	tyr	cys
C	leu	pro	his	arg
A	ile	thr	asN	ser/arg
G	val	ala	asp/glu	gly

Figure 46, theoretical two base genetic code

Example peptides from the two base code

ser arg ser arg   phe phe phe phe
UC AG UC AG   UU UU UU UU
AG UC AG UC   AA AA AA AA
val leu val leu   asN asNasN asN

ser leu asp tyr thr

UC CU GA UA AC

AG GA CU AU UG

asp arg ser tyr val

Figure 47 example peptides from two base code

But the three based code as noted by Crick had to have always been present. Perhaps in the following arrangement:

UCarg.CUasp.GAleu

serAG.leuGA.argCU

Figure 48 hybrid code using an amino acid in the third position

This last combination in fig 48 puts the amino acid in the third position and has it interacting with the nucleotide pair in the opposite strand. The interactions are my guesses but similar affinities based on shape complementarity are possible. This would have kept the third position in place making it easier to develop the three base code. It could also have been the basis for Woese's affinity based code development where stereochemical recognition between bases and amino acids, as paired across the strands, began the process of developing the genetic code.

Note that there are 16 combinations in the two base code which is more than enough to code for 10 amino acids plus some punctuations (start and stop codons, not shown).

Peptides encoded in the antisense strand of DNA have been predicted and found experimentally to bind to sense peptides (peptides encoded by the sense strand of DNA, the strand that is normally translated into proteins) with significant selectivity and affinity. Such sense-antisense peptide recognition has been observed in many systems, most often by detecting binding between immobilized and soluble interaction partners. Some receptors and their effector molecules have sense/antisense character in the binding sites. Data obtained so far on sequence and solvent dependence of interaction support a hydrophobic-hydrophilic (amphipathic) model of peptide recognition. (Tropsha A, Kizer JS & Chaiken IM. 1992 [39]).

Orgel has proposed that amino acids and RNA may have been at some prebiotic stage linked in a molecule or complex as part of the stabilisation of components of the RNA world [40]. His proposal is to link directly the anticodons with amino acids in a complex to make a crude genetic code. Eörs Szhathmary and John Maynard Smith have come up with a concept called "coding coenzyme handles" to cope with the mechanism for this [41]. A much simpler version of the code probably existed in the RNA world era and Ivanov has proposed a set of ten probable amino acids for the prebiotic era which he calls "ordinary"[42]. These gradually came to include the ten extra "special" amino acids at rates which he has calculated from the protein database. That would make the twenty we know today. But why those twenty as there are other amino acids not used in the twenty. These include ornithine, hydroxyproline and others all used in biochemical ways by cells but not included in proteins by the translation mechanism. The simpler ten amino acid usage fits nicely with a much simpler code Crick and others have proposed. Using a two base code there are 16 possible combinations for the ten amino acids leaving enough codons for the punctuation or stop codons plus some redundancies. Note that all of Ivanov's ordinary amino acids are coded for. However, it is important to note that as Crick points out there must always have been a three letter code of some sort.

Cycles of hot and cold (fire and ice) in a hot spring surrounded by ice and snow could produce a PCR effect, with melting and re-annealing of strands as they are copied. Another role for fire and ice.

PCR polymerase chain reaction

PCR is a method, used by molecular biologists, of amplifying DNA using a polymerase enzyme isolated from hot spring archaebacteria (Thermus aquaticus) known as Taq polymerase. The technique uses cycles of heating and cooling which allow alternately melting apart of DNA strands at high temperature, then cooling to allow copying of the two strands. This is followed by another round of heating then cooling with each round doubling the amount of DNA of the original strand. Primer sequences are used to restrict the copying to a section of DNA between the primers. Rounds of doubling very quickly result in massive amounts of the copied strand of DNA.

Cycles of hot and cold (Fire and Ice) in a snowbound geyser could produce a PCR (polymerase chain reaction used in laboratories to amplify DNA from a few strands to multiple copies). The process involves a 'melting' high temperature stage during which DNA strands separate, a copying phase where new copies are made and re annealed at lower temperatures, with melting and re-annealing of strands as they are copied. Another role for fire and ice.

Early life is likely to have been based on an iron/sulphur cycle and this may be seen today in the archaea, particularly those found around the black smokers and in thermal ponds associated with geysers or other volcanic zones.

Relevant to the discussion about the site for formation of the first organisms the thermal springs are sites where thermophilic archaea thrive. The colourful layers (Fig. 49) found in the hot springs are populated by archaea each of which occupies a different temperature or acidity zone.

Figure 49. Bands of colour reflect bands of archaea

Nucleoprotein networks

Nucleotide coenzymes have innate reactivity (Visser and Kellogg) and ribozymes show that catalysis by RNA is not only possible but it remains a common occurrence in modern biology. A list of modern biochemical reactions catalysed by ribozymes includes: see Fig.45: transpeptidation in ribosomes, ribonuclease A, Type II exon, tetrahymena ribozyme. However, the RNA world may never have existed in isolation, in fact it is my, and many others, contention that it must have existed in association with protective amino acids and peptides. There is significant support for this as detailed above.

Primitive selection of the fittest emerging through functional synergy in nucleopeptide networks

Anil Kumar Bandela 1, Nathaniel Wagner 1, Hava Sadihov 1, Sara Morales-Reina 2, Agata Chotera-Ouda 1, Kingshuk Basu 1, Rivka Cohen-Luria 1, Andrés de la Escosura 3 4, Gonen Ashkenasy 5

Proc Natl Acad Sci U S A 2021 Mar 2;118(9):e2015285118.

doi: 10.1073/pnas.2015285118.

Many fundamental cellular and viral functions, including replication and translation, involve complex ensembles hosting synergistic activity between nucleic acids and proteins/peptides. There is ample evidence indicating that the chemical precursors of both nucleic acids and peptides could be efficiently formed in the prebiotic environment. Yet, studies on nonenzymatic replication, a central mechanism driving early chemical evolution, have focused largely on the activity of each class of these molecules separately. We show here that short nucleopeptide chimeras can replicate through autocatalytic and cross-catalytic processes, governed synergistically by the hybridization of the nucleobase motifs and the assembly propensity of the peptide segments. Unequal assembly-dependent replication induces clear selectivity toward the formation of a certain species within small networks of complementary nucleopeptides. The selectivity pattern may be influenced and indeed maximized to the point of almost extinction of the weakest replicator when the system is studied far from equilibrium and manipulated through changes in the physical (flow) and chemical (template and inhibition) conditions. We postulate that similar processes may have led to the emergence of the first functional nucleic-acid-peptide assemblies prior to the origin of life. Furthermore, spontaneous formation of related replicating complexes could potentially mark the initiation point for information transfer and rapid progression in complexity within primitive environments, which would have facilitated the development of a variety of functions found in extant biological assemblies.

Conclusion: Stereospecific interactions between Nucleotides and peptides give rise to a crude code which develops into the genetic code.

BASIC REQUIREMENTS FOR LIFE

The simplest extant organism capable of reproduction is the bacterium. At one time these were very likely to have been the only organisms on Earth. However, any theory of how life began must also consider the origin and existence of other biological organisms not capable of self-replication. These are the virus and the prion. There may yet be other entities discovered in this group. I am including these entities because others who have considered the question of the origin of life have largely ignored them. They exist so they need to be explained.

The Origin and Evolution of Viruses as Molecular Organisms October 2009 Nature Precedings 4TY -

AU - Bandea, Claudiu Bandea proposes that viruses were part of the prebiotic evolution mechanism providing a means for certain reactions to be preformed which could not happen any other way.

Perhaps as Sherlock Holmes once remarked to Watson in "The Hound of the Baskervilles" "It is often the evidence which is difficult to explain which when properly considered can result in the complete elucidation of a case". Note that viruses have become very useful tools in the biochemist/molecular biologist tool kit and can be used for transferring genes. In fact, treatment of certain inborn errors of metabolism such as cystic fibrosis is now being performed using virus vectors to replace the defective gene.

Even the simplest self replicating organisms require a very complicated set of biochemical and other reactions to sustain them. These can be grouped as follows. There are primary metabolic reactions, which allow the organism to derive energy from ingested food. There are biosynthetic reactions, which allow the organism to produce a new set of components in order to make a copy of itself. Those components include the following: nucleotides for DNA and RNA synthesis; phospholipids for cell wall components, proteins of over a thousand types for all the enzyme catalysis and structural elements in the cell; the self replicating machinery (both RNA and protein) and the complex carbohydrates which are components of cell walls and cell surface glycoproteins. Even this very simplified list is made up of some very complicated molecules that would be very unlikely to have self-assembled. The complete list of molecules runs into the several thousands. To envisage a simpler molecule than this with any of the properties of self-reproduction is quite difficult.

Let us consider what the minimum requirements for a self-replicating system might be.

1. A template with a blueprint for making a new organism
2. A copying mechanism that is either error free or self correcting
3. A set of available raw materials from which to build the new entity
4. A way to prevent chaos from disrupting the process
5. A driving force to promote the creation of the new entity

There are the simpler entities, the viruses and prions I mentioned, however, which do not self-reproduce but which nevertheless carry information that can be reproduced. The virus does it by using the host machinery to make its components which are primarily DNA or RNA and protein for making a coat. Once inside the host the host proteolytic enzymes obligingly remove the virus coat. This exposes the naked DNA which can then either insert itself into the host DNA or can co-exist within the cell to be reproduced by the host machinery. The host machinery is then redirected from its original purpose by a series of virus-induced modifications, toward making the new DNA rather than its own and the new virus coat protein rather than its own components. The possibility exists therefore that early prebiotic reproduction of a DNA like molecule could also have occurred in some way. Perhaps a catalysis system existed separately from the 'organism' which gave rise to the organism components. A system of crystal genes, such as that proposed by Graham Cairns Smith, could give rise to such a scenario. The system that arose can be looked at as having come about in a series of steps that must be unbroken because they have continually built upon the original system to create ever increasing levels of complexity which can no longer resemble the original process.

Consider the analogy of manufacturing a television set or a computer. We could not possibly make one from scratch in our backyard garages even if we knew how to do so because we need the appropriate tools and raw materials. And even with the knowledge and the raw materials we would not be able to assemble the microchips and circuitry in the way that the factory would. The set we would make would lack the neat presentation features and probably a lot of the quality of image of a bought TV set. So there is a multiple set of layered steps leading up to the production of a TV set which must go back to the development of simple cathode ray tubes and other components before we can even begin to assemble the article we know today. Just in this same way the biochemical infrastructure that has gone into the production of multiple celled organisms has been assembled from processes which themselves were developed from other processes and so on back to the prebiotic processes that existed before life began. Newton once said that what he had done in his discoveries was to stand upon the shoulders of giants. So it is with the process of life itself.

Conclusion: Minimum requirements for a self-replicating system are considered.

THEORIES AND EXPERIMENTS OF MOLECULAR EVOLUTION

The evolution of the genetic code is a part of the theories of prebiotic evolution. Biotin theorists Visser and Kellogg referred to a 1967 paper of Hans Kuhn in which a theory was proposed whereby molecules could accumulate in porous rocks found in tidal pools. The molecules would be replenished by flushing only at high tide, which occurs once every month, thereby giving the intervening twenty seven days for the pool to evaporate slowly and concentrate its contents. The molecules in the rocky pores were protected from UV damage during the concentration. The rocky surface could contain the kind of crystalline template suggested by Cairns-Smith so that the slow concentration could cause crystallisation of organic molecules. The molecules in question could have been amino acids, nucleic acids or something resembling them. A common precursor for both types of polymer could have been formed with imidazole as the base or amino acid and a simple sugar such as ethylene diol (glycol) or perhaps glycerol. Some pools or crevices may have contained different molecules allowing the formation of some polymers as the crystalline arrays condensed. In a famous set of experiments it has been shown by Stanley Miller that lightning passing a mixture of gases that reflects the prebiotic atmosphere can produce amino acids and other biochemically interesting products. In Miller's experiment with reducing atmosphere gases and a spark, imidazole and purine nucleotides were readily made from cyanide. These experiments have been dismissed of late as the gases involved are now thought to be different from those which Miller used. It is also possible to generate many simple sugars from spontaneous condensation of formaldehyde, another component of the early atmosphere. The biggest criticism of these types of experiment is that the end product is always some intractable goo from which it is almost impossible to envisage a spontaneous re-organisation of any type to result in an ordered polymer. Entropy rules in this type of experiment. My conclusion is that these experiments obviously lack something in their design which would allow the formation of useful molecules. Carl Sagan has shown the presence of small biological precursor molecules and even amino acids in the atomic adsorption signatures of stars indicating their presence in the universe. However they arose, it seems certain that polymers of some sort did form perhaps from cycles of hot / cold, wet / dry and freezing/thawing far more extreme than today's greenhouse gas-controlled even temperatures. Such conditions can be found in the icy waters of modern Iceland where volcanic lava flows into the ocean from volcanic activity. The proximity of extremes of temperature is ideal for cycling reactions to occur. Temperature cycling is the very fundamental of the PCR reaction which biochemists use to amplify copies of DNA for analysis and which incidentally is made possible because of an extremophile-derived enzyme isolated from archaea bacteria found in hot springs and which can survive up to 95 degrees. Perhaps a chemical version of this reaction, catalysed by a microenvironment about which we can only speculate (see later), was responsible for a magnification in amount. Thus it is possible that a polymer of say imidazole (which has structural similarities to both amino acids and pyrimidine bases, Fig. 50 could have formed and made a template onto which other similar pyrimidines could assemble.

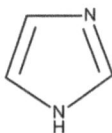

Figure 50 Imidazole

Such a polymer would be written in molecular shorthand as I-I-I-I-I-I-I-I-I-I-I-I-I-I. The cycles of hot and cold would then promote amplification of the molecules so that a number of polymers would have been formed. Another consequence of the hot springs as an origin is that reactions occur much more rapidly at high temperatures thus allowing the sorting reactions in the chemical evolution process to occur very rapidly. With this ability to replicate, these polymers would have increased to concentrations above others in the soup. Polymers with inclusions of different pyrimidines could have occurred, formed on crystal templates in which regular flaws in the crystal in the form of metal chelates occurs. An imperfection or a substituted metal ion in the crystal would have favoured the deposition of a different pyrimidine in that position in the polymer. For example, in our imidazole polymer we could get an occasional uracil but on a regular basis. For example, I-I-I-U-I-I-I-U-I-I-I-U-I-I-I-U-I-I-I. Orgel [40] has reviewed a number of studies in which molecules can replicate without the aid of enzymes. This probably took a long time to develop into the RNA world that Walter Gilbert and others propose. So, with the time scale of 200 myr to work with, we need to see how all of the modern biochemical reactions were put in place.

What set up the conditions for life to arise?

The force of evolution at its most primal can be described as the survival of genes; 'selfish genes' as Richard Dawkins has described them. This is a reflection or more correctly a remnant of the conditions that prevailed to allow life to arise. The copying of genetic material to make more of itself must have had its origins in early molecules copying themselves or being replicated; to what purpose? There can have been no better force than numbers or mass. Why? Because it can. As Ernst Mayr points out in his book "What evolution is" the whole point of trying to find a driving force for evolution is answered by the observation "because it can". If a new development can occur eg. to fill an empty niche, then it is more than likely that it will. The fact of increasing populations of the same molecule would be force enough since those numbers would mean the production of a concentration of a substance which was able to survive longer because of its mass than other molecules under forces of destruction. The mass of molecules would have only the driving force of like rather than unlike to promote self-formation against an almost overwhelming power of destruction. What were those forces of destruction? The most damaging on the planet surface would have been ultraviolet radiation, which would have been hitting the Earth unchecked without atmospheric ozone derived from oxygen to trap it. However, life probably began underwater where UV does not penetrate. These UV levels could have one day returned with the complete destruction of the protective ozone layer until international cooperation reduced the use of CFC molecules responsible for the ozone depletion. We know UV radiation causes mutations in DNA even at today's reduced levels, causing melanomas. The protection against UV damage would have been of primary importance once the complex molecules came to the planet surface but if they were formed in the deep oceans near black smokers the sheer mechanical forces caused by high temperatures would have pressured the survival of molecules. Polymers that survived in this environment would have had both thermal and chemical stability while being able to promote self-replication. Perhaps the conditions were similar to those afforded by a cell membrane where there were two or more interacting polymers in a site which promoted the polymerisation of each on surfaces which were physically very close to each other.

Once on the surface those molecules, which would have survived longer in this environment, would have had some protective advantage. That could readily have come from an association of RNA-like molecules with amino acids or peptides and/or other hydrophobic molecules. Those peptides with the highest potential for binding to DNA are the aromatic ones because of their hydrophobic interaction capabilities. These amino acids coincidentally have UV absorption properties and can therefore, by acting as shields, protect nucleotides from UV destruction. Thus, the first association of nucleotides with proteins was probably a successful cooperation to allow the nucleotides and pyrimidine polymers to survive longer in a destructive environment. Now let us imagine that the polymer is not just imidazole but a repeat sequence of pyrimidines such as Imidazole-uracil-cytosine or in molecular shorthand IUC. A repeat sequence of this would read I-U-C-I-U-C-I-U-C-I-U-C. Now let us look at amino acids that might bind to and protect this sequence from UV damage. The criteria for the most successful protection would be the tightest binding from a mixture of potential candidate amino acids. It might even have been short peptides formed by the same self-assembling, crystal driven process. Now we have a potential copolymer which can survive UV damage. Where does that get us? The proto-RNA that is forming has some very weak and not very specific catalytic properties. It can cause breaks in other polymers to occur at certain weak points in the polymer. This has the effect of damaging those proto RNA molecules that possess a certain sequence which can be cleaved more readily by UV. This means that some sorting of molecules can occur. It might mean that some polymer sequences become even more enriched while others become degraded and their components reused to make more of the dominant polymers. This is still random but by the destructive force of UV damage as an external impetus the molecules that survive have unique survival properties and part of this is the specific association with peptides made up of aromatic amino acids. As time goes by they are the polymeric molecules that can promote their own survival by the mechanism I have described.

The catalytic property I described has been shown in certain types of RNA called hammerhead ribozymes which act to split themselves at a particular position in the sequence (see above). Just as likely is that RNA molecules developed which could promote the formation of specific bonds. One such bond between RNA and an amino acid has been demonstrated by a simple laboratory-made mutation of the ribozyme found in Tetrahymena. If this simple relationship between amino acids and polymeric nucleotides has been arrived at in this set of logical and very plausible steps and you are still with me then the next steps are more spectacular but equally plausible.

A study of the interaction of glycine and its oligohomopeptides with formaldehyde and acetaldehyde under possible primitive Earth conditions Orig Life

. 1983 Dec;13(2):97-108. C P Ivanov, O C Ivanov It is established that glycine and glycine oligohomopeptides interact with formaldehyde and acetaldehyde in a homogeneous weak acid medium (pH 3.3-3.7) at mild temperatures (60-80 degrees C) in the absence of inorganic solid substances. Together with the expected serine and threonine, the formation of alanine, glutamic and aspartic acid, norvaline and isoleucine, as well as four non-protein amino acids is also established. It is suggested that the non-protein amino acids are hydroxymethylserine, hydroxymethylthreonine, hydroxymethylaspartic acid and gamma-amino-delta-hydroxyvaleric acid. The modes of formation of all protein and non-protein amino acids are discussed. These results strengthen the probability that similar processes may have been one of the pathways for the prebiotic synthesis of amino acids on primitive Earth.

It would have been too complicated to develop from a two letter code and therefore unlikely to reorganise completely from one code to the other. This can be accommodated if a two-letter code contains a third unused position or even more likely a position occupied by a non-nucleotide filling the position, such as an amino acid or even a dipeptide (see fig 48).

Conclusion: Further development of the genetic code

Chapter 6

SENSE-ANTISENSE PEPTIDE RECOGNITION, PEPTIDES AND PROTEIN FOLDING

Antisense-sense affinity

Peptides encoded in the antisense strand of DNA have been predicted and found experimentally to bind to sense peptides (peptides encoded by the sense strand of DNA, the strand that is normally translated into proteins) with significant selectivity and affinity. Such sense--antisense peptide recognition has been observed in many systems, most often by detecting binding between immobilized and soluble interaction partners. Data obtained so far on sequence and solvent dependence of interaction support a hydrophobic-hydrophilic (amphipathic) model of peptide recognition. (Tropsha A, Kizer JS & Chaiken IM. 1992 [39]). These authors speculate that the antisense peptide –peptide recognition reflects the DNA-antisense DNA strand recognition by using the same shape complementarity in both cases. In other words the peptide-DNA recognition uses and reflects the same system that gave rise to the genetic code where peptide -DNA binding was based mainly on shape complementarity. This must have been fairly crude and error prone and was eventually replaced by the much more efficient tRNA mediated ribosome catalysed assembly of peptides that occurs in modern biochemistry. While this modern system is highly organized and complicated it is hard to see how it could have arisen spontaneously. Not only is the ribosome a large and highly organized cell component consisting of both a catalytic RNA and many surrounding protein subunits to hold it all together but it has to recognize the incoming tRNA molecules and the messenger RNA being translated into protein. Each of the associations of transfer RNA with its specific amino acid relies on individual enzymes, amino acid tRNA synthetases to connect the correct amino acid to its correct carrier tRNA molecule. A simpler but cruder system such as the peptide –DNA shape recognition system could well have been the precursor to all of the above. It remains to fill in the steps in between but as pointed out in a cold Spring Harbor Symposium the new system, once established would have taken over because of its reliability and efficiency leaving very little of the original system to recognize today.

The sense-antisense observation turns out to be more intriguing than just recognition. Chaiken and co-authors looked at the complete set of amino acid sense-anti sense interchanges an interesting pattern emerges (fig 3.) These authors make the following observations: 1) the sense-antisense exchanges break into three non-communicating groups. Since the grouping is the result of the specific redundancy of the genetic code it is probably not accidental. 2) The exchange requires the genetic code to be complete since only four exchanges are duplicated. The specific redundancy of the code is necessary for most of the exchanges. Eg. isoleucine has three codons and exchanges to three different amino acids while valine has four codons and exchanges to four different amino acids. 3), When the mitochondrial code is used the pattern is retained and only four codons are redundant. Since the mitochondrial code is probably older and probably arose earlier in evolution it raises the possibility that the sense-antisense exchange pattern and the redundancy of the genetic code may be evolutionarily related.

There has been some effort in the literature to find evidence for or against this observation and there are specific examples of this recognition e.g. the use made by Knutson to identify the site of interaction of insulin and its receptor [43]. However, there have been some equally unsuccessful examples such as an attempt to find a regulatory peptide for angiotensin II.

This sense-antisense affinity relationship defies logic in many ways because the genetic code is set up so that the antisense peptides have the opposite hydropathicity to the sense peptides. E.g. where a hydrophobic stretch of amino acids occurs on one peptide the antisense peptide will have a hydrophilic character, so why should they recognise each other? The answer seems to lie in folding. Each of these peptides seems to have a complementary shape to the other. This is a more fundamental observation than it seems. Shape complementarity is a common theme throughout biochemical reactions in all cells. The interaction of many proteins and nucleic acids is based first on shape complementarity then only secondarily on charge or other interactions. One of the most important interactions in the cell signalling pathway is that between the components of the receptor complexes and their hormone agonists. All enzyme reactions are absolutely dependent on the shape complementarity between the enzyme and its substrate. There it is refined by allowing subtle changes in the active site which act to bend the substrate into a shape resembling the products to be formed. In this way the enzyme 'approximates' the state which the substrate has to reach to form products which is called the transition state. Even more obviously, if two molecules do not have complementary shapes they cannot get close enough to each other to interact successfully. Most importantly of all perhaps is the folding of a protein with no voids. Any protein which folds incorrectly leaving voids within its structure is unstable and becomes readily degraded within the cell. The stability is based on the strength of hydrophobic interactions between amino acids within the protein structure. The centre of a protein is described as being like an oil drop in its hydrophobic nature while the external surface has most of the charge and water interacting qualities necessary to make the protein both soluble in water and able to interact with other proteins and with metabolites.

The best clues as to how the code arose come from the sense-antisense peptide recognition phenomenon. Chaiken ref [39] and his group describe the relationship in their landmark paper entitled "Making sense from antisense" in which they describe the shape complementarity argument. If the peptide made from a strand of DNA is mixed with that same DNA strand the two will bind together. Note that the association is between the sense strand and the peptide, not the antisense strand. This is important later. This observation leads to the further observation that if the other strand of DNA, the antisense strand, is also translated into its antisense peptide then the peptide strands made from the complementary DNA strands will bind to each other. This proves to be a real phenomenon demonstrated by the ability to make affinity matrices of one peptide which can be used to select out the antisense peptide from mixtures of peptides. The interaction of the two DNA strands is achieved by a series of hydrogen bonds making the DNA-DNA interaction very strong. However, the DNA-peptide interaction is weaker because it is only driven by shape complementarity and partially by hydrophobic forces. (See Fig. 51 for diagram).

Antisense peptide recognition

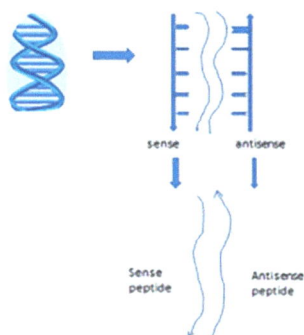

Figure 51. DNA helix-> unfolded to show DNA bound peptides->
isolated sense peptide binds to antisense peptide.

This is the basis of a theory, known as the stereochemical theory, first proposed by Carl Woese in 1966 [38]. Woese, a microbiologist was involved from the beginning in working out the genetic code. The shape complementarity of amino acids and their codons is no trivial explanation, as it becomes a central theme in biochemical interactions everywhere. Enzymes operate by a mechanism once described as lock and key where the compound (substrate) being acted upon by the enzyme fits into the active site like a key in a lock. The binding is so tight that it allows enzymes to distinguish between many like compounds and pick out the one that is to be acted upon mainly by its fit. Similarly, antibodies owe their specific action to the ability to form many different shapes that will bind to any foreign substance. Once one antibody finds a strong enough fit, many more copies are made which then all bind to the substance to take it out of action. Many other biochemical reactions such as cell signalling rely on a close affinity, or lock and key interaction, of two components which form part of a relay of events leading to such things as development of a limb in the embryo or the regulation of pathways of metabolism.

PROTEIN FOLDING

Protein folding, a problem often regarded as the holy grail, the main remaining problem to be solved in protein chemistry, is essentially the problem of how a linear sequence of amino acids will always fold to form the same tertiary structure. What directs that final shape? It must be from the information contained in that linear sequence but attempts to predict the final shape have failed until recently. The puzzle has been approached with a powerful computer algorithm devised by Steven Mayo in which shape complementarity is one of the many variables considered. See video presentation: https://www. ibiology.org/archive/protein-design/#part-1

Protein folding is important for the origin of LUCA for the explanation of how early peptides could develop catalytic surfaces as Kauffman predicts. It is the capacity for peptides to fold into shapes as they are produced that has challenged protein chemists trying to predict what determines the folds and why they always have the same outcome from the same sequence of amino acids.

Predicting protein tertiary structure has been the most elusive puzzle of recent times to protein chemistry. The hypothesis of a thermodynamic driving force for this process was developed by Anfinson (1973) [44] but other work suggests a kinetically driven process is just as likely. The tertiary structures obtained from X-ray diffraction of crystals are available for a growing number of proteins but there are many more for which this approach either does not work, because not all proteins will crystallise, or is inapplicable because pure protein cannot be obtained in sufficient quantities. For these proteins it would often be of value to predetermine their likely folding conformation. The problem is therefore to predict structure for these proteins from their linear amino acid sequence. There are many theories as to how folding proceeds and the translation of theory into practice has been attempted by many with the most widely known being the Chou and Fasman prediction (1974) [45]. This results in a crude prediction of domains folding as alpha helix, beta pleated sheet or "random coil". Other approaches (Kyte and Doolittle [46]) have used hydropathic plots to predict physical interactions. None of these approaches has been wholly successful and the most successful practical approach has been to base prediction on similarity with an already known structure from a member of the same family of proteins, ie one with known sequence similarity to the unknown protein and for which there is known tertiary structural information

All of these approaches seem to be missing some vital feature for they cannot predict structure from the initial sequence and yet the protein always folds the same way every time. A very recent announcement in 2020 by a company called Deep Mind has used an AI machine to predict protein structures with a high success rate. This is the same AI machine that DeepMind used to win a game of Go, a notoriously difficult board game, against the world's best human player. However, having an AI machine predict structure does not tell us how it is done or what the underlying principles are, but it is a promising first step.

An extra player in this field is the discovery of chaperonins which are proteins which help other proteins to fold correctly. Not all proteins require them but many do, especially it seems, proteins made in the endoplasmic reticulum destined for export from the cell or for an organellar location. In any case the role of chaperonins seems to be to prevent improperly folded proteins from aggregation (Martin and Hartl 1993), although other roles are possible such as to squeeze improperly folded proteins into their final shape. In either case this function has a kinetic, rather than thermodynamic, role. Therefore part of the need for chaperonins may lie in the existence of more than one folding solution for proteins. This extra process only displaces the final structure by one extra intermediate, however, and does not alter the fundamental nature of the problem. In addition, some attention has been given to pro-regions in proteins. That is the regions of proteins when synthesized as larger polypeptides are later removed to form the mature protein. It seems that pro-regions may direct proper folding of proteins and once this function is fulfilled they are removed. Proteins with pro-regions are therefore of interest to study for protein folding processes. Insulin has a small pro-region and the protein thus synthesized contains the three disulphide linkages that hold the two peptide structure together once the pro peptide is removed. Note that prions exploit the role of folding templates, causing proteins to adopt a folding shape which is not the correct functional shape but a non-functional, disease causing one. Jakob Cruetzfeld disease is an example.

The solution to the folding puzzle probably lies in an as yet unexplored area where parameters previously unaccounted for are at work. This proposal seeks to test that one such parameter is shape complementarity. The central premise is that overlying the first level folding into domains with alpha helical or beta sheet structures which may depend to some extent on the electronic or physical interaction of domains due to their charged or hydrophilic/hydrophobic character, there is a level which requires shape complementarity to complete the folding process. Put simply it means that if proteins physically will not fit together then they will not fold properly. If proven this will imply that no folding prediction theory or protocol that has yet been tried will work 100% unless it takes into account the final shape of the interacting domains. Unless the predictive method shows that interacting domains will fit together in a completely complementary way the rest of the prediction is useless. One extensive recent review of protein structure prediction (Eisenhaber et al. 1995 [47]) shows that the closest this field has come to such an approach is to predict conformation space. That is to predict the shape which a given peptide will occupy, which is difficult because there are too many parameters. The difference is that the shape complementarity approach fits micro shapes together, thus reducing the number of variables to only a few possibilities which then determines the macro shape of the domain or region. This approach would build on other well established techniques such as secondary structure prediction and "threading" to develop a picture of the whole protein. Clinical application of this work may be found in elucidation of the aetiology of the prion diseases of Bovine spongiform encepalophathy, Jacob-Cruetzfeld in which proteins are apparently induced to take up alternative conformations by copying a misfolded initiating prion, due to some disease-related stimulus, which they do not adopt in a healthy individual.

The folding problem

A simple game analogy to the folding problem is the computer game called Tetris in which a series of pentominoes, or shapes made from every permutation of five cubes linked together (there are twelve possible shapes derived by this process). The shapes are placed on a grid randomly and sequentially so that a solid rectilinear volume is filled with **no voids**. Incorrectly placed pieces result in larger and larger voids until the player runs out of room and the game terminates. This is very analogous to what proteins do in their inner surfaces where amino acids fit together so well that there is no room for even water molecules to enter (except where they are wanted). It is therefore possible to use the logic inherent in the programming of the Tetris game be applied to the three dimensional shapes of the twenty amino acids to generate algorithms which can simulate protein folding.

There would of course be an extra variable in that amino acids can adopt a number of shapes in solution and are flexible to some extent, but this is a fine tuning matter for later. The initial approach would be to record the dimensions of the most extended known solution structure of each amino acid and to develop programmes to begin to fit those shapes together in three dimensional space. The programmes would develop three dimensional images of amino acids tethered in a string, just as a sequence is encountered in a peptide. One simple model will then be developed of two such peptides interacting. The antisense peptide interaction model of Tropsha et al. (1992) could be used as a model system to test the interaction of peptides. This model shows that peptides derived from the sense and antisense strands of a DNA sequence recognise and bind to each other by shape complementarity. This therefore models the simplest level of interaction allowing the central hypothesis of this model to be tested. The antisense peptide model can be taken a step further because it has been used to identify the shape of the surface which interacts with some peptide hormones. This data can therefore be used to test this shape predictive approach for these same hormone -receptor interactive surfaces. Once the system has been developed at a simple peptide level it is proposed to move on to a small protein. Proinsulin will be ideal for this purpose as there is a considerable amount of tertiary structure data available from X-ray crystallographic studies. Once tested the method can be applied to the interior regions of known proteins where the interaction is well characterised. There is also data on the tertiary structure and folding of many other proteins and this will be built upon as the project progresses to the more complicated proteins.

IS THERE A PLACE FOR ANTISENSE IN EVOLUTION ?

Protein folding and the genetic code may ultimately be linked. My proposal is that specific interactions between peptide pairs which are now seen as sense-antisense pairs arose in the prebiotic milieu as much less specific complexes with specific RNA duplexes in order to stabilise and protect those duplexes against the ravages of UV damage and free radical unzipping. Such a scenario would have occurred in the prebiotic era. In fact this is the likely driving force for the evolution of the prebiotic molecules. To survive in the environment of such destructive forces meant that molecules would need to be able to be present in force of numbers. Molecules that did not do this would become part of the disorganised soup. This does not mean that there was any purpose or direction to the prebiotic era. Just that in the destructive climate the fact that some molecules were able to survive if they perpetuated themselves resulted in their increase against a chaotic trend. Mass action or safety in numbers. This self-perpetuation against the destructive milieu has set the pattern for evolution where the selfish gene, or DNA itself has found infinite ways to survive in its many organism forms.

A paper by Hendry et al. [48] shows that there can be specific amino acid associations with specific nucleotide pairs and that these associations are determined largely by shape. This is strong support for the stereochemical theory. They propose a combination triplet codon of two bases and one amino acid. These DNA-amino acid complexes could have promoted peptide bond formation resulting in peptides with antisense recognition which also had shape complementarity by virtue of their formation from a shape template. I further propose that the single strands of those duplexes had replicative activity of some kind and that this constituted the first kind of self replicating system. This later evolved into both the genetic code and into the hidden code which carries the prerequisites for protein folding that we see today.

Zull et al TIBS July 1990, p257-261 also thought of the folding aspect and went looking for complementary sequences [49]. They compared the frequency of naturally occurring complementary sequences , or palindromes, to those generated randomly and found matching numbers which suggests that complementary sequences in modern proteins do not occur as a specific mechanism for the initiation of protein folding. However, a Japanese group did find complementary sequences in specific proteins. Others have proposed specific stereochemical recognition mechanisms. However, what they did not show, are places in crystal structures of proteins where the folding brings those "antisense peptides" coded for by the DNA sequence complementarity together. Now this may be an ancient mechanism which has been superseded by random incorporation of mutations over time and they did not look for corrupted complementarity or for a simpler code that may have had only two bases. Primitive organisms such as archaea and bacteria may not have the clues any more as they have gone through many more replicative cycles than higher organisms. In fact the Archaea and bacteria split from each other at the same time and differ in features such as the way they have used cell membranes. But this is not a reflection of their ancient origins, more a feature of different body plans being built upon as they evolved separately.

Woese stereochemical theory and antisense peptide recognition:

A conclusion from Chaiken et al was that the phenomenon of peptide-peptide binding is a reflection of the DNA strand/peptide binding which gave rise to the genetic code. The argument goes that DNA/DNA binding is made up of tight hydrogen bonds across the two opposing DNA strands but take one DNA strand away and the 3D structure remaining will bind to the peptide that the strand makes in protein synthesis in the ribosome. Now replace the DNA strand with the antisense peptide strand and again stereochemical recognition occurs resulting in peptide-peptide binding. Woese' hypothesis is essentially that the middle step, the DNA-strand peptide recognition is how the whole process began and how the genetic code arose. See fig 51.

Implications for modern biochemistry

The place to look for information may be in proteins which provided the breakthrough transition from the RNA world. The aminoacyl tRNA synthetases are being looked as these bring together RNA and amino acids in a highly specific way. Steitz calls this group of enzymes the second genetic code.

It is the specificity of these enzymes upon which the fidelity of translation of mRNA into proteins depends. The correct translation of messenger RNA into protein requires that the correct amino acid is first attached to the correct transfer RNA molecule. It turns out that recognition is at another part of the tRNA molecule than the anticodon region. That means that the whole tRNA molecule is important in the process. Therefore an incorrectly folded molecule will not be correctly aminoacylated.

Other modern day uses of this kind of esoteric research are many and may include an understanding of protein folding and being able to predict it and an understanding of exon splicing mechanisms and its implication for mutations in that area. Ultimately it will lead to the underlying principles behind biochemistry and the phenomenon we call life.

Conclusion: Sense-antisense peptide recognition is a reflection of the stereospecific interaction of peptides with the DNA that encodes them.

Chapter 7

SUMMARY

SUMMARY OF THE STAGES ON THE PROCESS TOWARD THE ORIGIN OF LIFE.

STAGE 1

Vesicle formation in fire and ice.

Encapsulation of amino acids in vesicles (Luisi) (overcoming entropy).

Ten prebiotic amino acids.

Catalysis of peptide formation, clay and/or sulphuric acid (Crystal theory of Cairns-Smith).

Peptides emerge that self catalyse (Kauffman).

Peptides "evolve" to develop other catalytic functions.

Energetics provided by sulphur chemistry (De Duve). Iron/sulphur energetics.

STAGE 2

Incorporation of bases into peptide complexes.

Symbiotic relationship between amino acids purines and pyrimidines of mutual benefit.

Catalysis of RNA formation-incorporation of polyphosphates.

Catalytic RNA (Altman, Cech, Symons).

RNA world emerges (but not in isolation) (Crick, Gilbert). The RNA-peptide world.

Fire and ice give PCR-like conditions.

Crude genetic code arises based initially on ten prebiotic amino acids (Woese, Ivanov).

Two base code (Crick)

Other nucleotide functions (Visser & Kellogg).

STAGE 3

The crude genetic code produces more variety of catalytic peptides and catalytic RNA.

These eventually create tRNA molecules and DNA.

Evolves into three base code.

supporting metabolic processes develop.

Voila LUCA.

Delivery of water and a variety of organic compounds to the early-formed Earth via comet and asteroid impacts supplied the Earth with raw materials, organic compounds, water and fatty compounds. Plate tectonic activity generated hot springs which provided chemical drives for vesicle formation, temperature gradients, salt gradients and proton gradients across membranes. In hot springs with ice and snow a *fire and ice* environment is created. Incorporation of starting materials such as amino acids, sugars and purine/pyrimidine bases in vesicles along with clay templates provided a means for peptide formation in a dehydrating, possibly acidic, medium.

Vesicle fusion and mixing of contents creates an environment where peptides grow among other chemicals to eventually form catalytic peptides that promote their own formation. Energetics are provided by polyphosphate hydrolysis or sulphur chemistry or proton gradients. Out of this comes activities which will promote the formation of new bonds between compounds which make new compounds and new peptides with activity. Promoted by an association of molecules with affinity for each other and by protection against damaging UV, new bonds are made between sugars and polyphosphate and/or sugars and purine or pyrimidine bases to form nucleosides and then nucleotides.

Polymerisation of nucleotides on the clay template forms something like RNA which develops catalytic surfaces and a whole new metabolism or RNA world.

This RNA is not naked but has a strong and codependent association with protective peptides and other organics. Stereospecific associations lead to a two base code which promotes the use of ten amino acids in peptides. The two bases have a third position occupied by a peptide or an amino acid which cross pairs with a dinucleotide on the opposite strand. A third base is eventually brought in but a whole array of other helper molecules are developing by then including transfer RNAs, Ribosomal RNAs and possibly even DNA. It should be noted that there are many modern biochemistry systems where there is an intimate association between peptides and nucleotides or RNA or DNA. In particular the complex of the ribosome is an intimate association of peptides with the RNA that catalyses the peptidyl transferase reaction to make peptides from messenger RNA. Also the association of transfer RNA molecules with the enzymes that couple them to their respective amino acids require a close binding association with the tRNA and indeed must recognise the particular tRNA molecule to which the individual amin acid is to be attached. Other associations include DNA polymerase, DNA transcriptase and many others.

Once all this is a working system we have LUCA.

This may not be the way it happened but in telling it as a logical sequence it provides a plausible pathway on which to work. More and more original research is coming out that adds more clues to this puzzle and I have tried to keep up with that although it is a moving target. The main thing is that it should provide aspects for an enquiring mind to want to find out more.

Charles Darwin did not know about DNA but if he did, he would undoubtedly have made some contributions more insightful than the "warm little pond" concept.

APPENDIX

<u>The processes of life</u>

<u>A very good array of videos of various living cell processes is available on youtube. Caution that these videos can be taken down at any time and a good biochemical text can be consulted instead. Eg Alberts et al. The Molecular biology of the cell</u>

<u>The copying of DNA in a cell: https://www.youtube.com/watch?v=Qqe4thU-os8</u>

<u>Protein synthesis: https://www.youtube.com/watch?v=NDIJexTT9j0</u>

<u>There are many more youtube videos worth watching to explain the wonder of cellular function at a molecular level.</u>

DNA

The information to make a cell is contained in a molecule called deoxyribonucleic acid or DNA for short. DNA is a stable molecule which exists in the form of a double helix ie two helices twisted around each other. This molecule is very long and can contain thousands of small regions which contain the code for making proteins. There are approximately 30,000 genes in the human genome encoded for by our DNA. The code is in the form of four nucleotide molecules known as bases. They are known by the shorthand of A, G, C and T which are known chemically as Adenine, Guanine, Cytosine and Thymine. The Double helix is a stable structure whose composition is such that base pairs are made by hydrogen bonding between the bases and these are what holds the two helices together. The base pairing is the crucial factor that allows DNA to be copied to make exact replicas of itself in the process of cell division. The DNA is copied by complex enzyme controlled mechanisms to make more DNA containing the same sequences. This cell division is a process called mitosis which allows the cell to reproduce. The accuracy of copying depends upon the base pairing which occurs between strands of DNA. This base pairing has A always pairing with T and G always pairing with C. The copying is performed by a set of enzymes specific for this purpose, the main one being DNA polymerase. Note that only about 2% of human genomic DNA is translated into the 30000 or so functional proteins. The remaining 98% is of as yet unknown function but there are strong clues that it directs the cell to make a subset of possible proteins as it would be unproductive for every cell to make all 30000 proteins. The process of switching off the translation of unwanted proteins involves the methylation (silencing) of genes such that they are no longer accessible to the transcription machinery.

DNA transcription into RNA (https://www.youtube.com/watch?v=_Zyb8bpGMR0)

The process of making proteins from DNA is very complex. DNA must first be split apart and short sections (genes) of one strand is copied into a special variety of RNA called messenger RNA (mRNA) from genes which are short sections of DNA defining a protein product. The strand which is translated into proteins is called the sense strand. This is genomic DNA converted into RNA which must then be trimmed to remove introns this requires precise cutting and rejoining of the RNA strand which then becomes messenger RNA.

RNA translation into protein)

The messenger RNA delivers the gene sequence to the translation machinery known as the ribosome. This is a very large complex of proteins and RNA molecules, assembled together which functions specifically to make proteins from messenger RNA.

The complex translation machinery brings together the messenger RNA, which sits in a special groove on the ribosome, and individual amino acids especially activated for joining together with other amino acids. The amino acids are coupled with another type of RNA molecule known as transfer RNA. Each amino acid has been coupled to its own unique complementary transfer RNA by a unique set of enzymes (amino acid-tRNA synthetases) designed to link these two components specifically together. Therefore there are at least 20 individual transfer RNA types. Without the unique specificity of these enzymes, the transfer RNA-amino acid synthetases, the process would not work. Each transfer RNA has a unique three base anticodon region which reads the codon region on the messenger RNA lying in the groove on the ribosome. Once the recognition has occurred the amino acid is coupled to the previous amino acid by another enzyme which sits in the groove on the ribosome and is known as peptidyl transferase. The transfer RNA molecule is then released and the protein has grown by one amino acid. This complicated process is fairly rapid and ends when the ribosome reaches a termination signal which occurs at the end of the messenger RNA molecule and the peptide or protein is then released from the ribosome. The protein folds a domain at a time as it is produced and does not wait for the entire sequence to come out before doing so.

The Genetic code

The genetic code is the set of interactions between the DNA gene sequence that allows that gene sequence to be translated into proteins by the above process. There are four DNA bases in the DNA sequence and three bases in a codon allowing a matrix of 4x4x4 codons or 64 possible codons which code for the 20 known amino acids as well as a start signal and some stop signals. There is considerable duplication in the code because 64 codons code for only 20 amino acids. This is known as redundancy but it turns out that the number of times an amino acid is used in a protein is reflected in the number of different codons that code for that amino acid. Eg. leucine is the most frequently used amino acid of all and it has the most codons (6) coding for it.

The triplet codon is 'read' by a molecule called transfer RNA (tRNA) which has a triplet anticodon region which interacts with and 'reads' the DNA sequence. All of this is performed within a very large molecular machine called a ribosome. One of the most important reactions which requires absolute integrity is the coupling of the tRNA molecule to its specific amino acid. This is accomplished by a set of individual enzymes called aminoacyl tRNA synthetases. It is crucial that these do their specific jobs with no mistakes in order to allow proteins to be correctly made.

REFERENCES

1. Life on Earth, Attenborough D, 1979, Collins/BBC.
2. On the Origin of Species, Darwin C, 1859.
3. The Ancestor's Tale, Dawkins R, Weidenfield and Nicholson, London
4. Wonderful Life, Gould SJ, 1989 Penguin
5. Life on a Young Planet, The first three billion years of evolution on Earth (2003) Andrew Knoll, Princeton University Press
6. Frontiers of Astronomy, Hoyle F 1955,
7. Szathmary E, Proc Nat Acad Sci USA, 90, 9916-9920 (1993)
8. Wallace, M.W.; Gostin, V.A.; Keays, R.R. (1989). "Geological Note: Discovery of the acraman impact ejecta blanket in the officer basin and its stratigraphic significance" (PDF). *Australian Journal of Earth Sciences.* **36** (4): 585–587.
9. Comets and the Origin and Evolution of Life. 1st Edition (1997) Eds. Thomas PJ, Chyba CF, McKay CP. Springer-Verlag, New York.
10. Comets and the Origin and Evolution of Life. 2nd Edition (2006) Eds. Thomas PJ, Hicks RD, Chyba CF, McKay CP. Springer – Verlag , New York.
11. J. Oro and AP Kimball, Arch. Biochem. Biophys. 94, 217 (1961)
12. SL Miller - Cosmochemical Evolution and the Origins of Life, 1974 - Springer

13. Saskia Lamour,[1,2,†] Sebastian Pallmann,[1,†] Maren Haas,[1,2] and Oliver Trapp[1,] Life (Basel). 2019 Jun; 9(2): 52. prebiotic Sugar Formation Under Nonaqueous Conditions and Mechanochemical Acceleration

14. Biogeochemistry, an analysis of global change. 2nd edn. by Schlesinger WH, (1997) Academic press

15. Origins of Life 2nd Edition (1999) Freeman Dyson Cambridge University Press

16. Higgs PG, Chemical Evolution and the Evolutionary definition of Life, J Mol Evol 2017 Jun;84(5-6):225-235

17. Tessera M, Is pre-Darwinian evolution plausible? Biol direct 2018 Sep 21:13(1):18

18. Kunnev D Origin of Life: the point of no return. Life (2020) Nov 3:10(11):269

19. Visser CM and Kellogg RM, (1977) Bioorganic chemistry 6,79-88

20. Visser CM and Kellogg RM, (1978) J Mol. Evol 11, 163-169,

21. De Duve C, Vital Dust, 1995 Basic Books, New York

22. AG Cairns-Smith, Seven Clues to the Origin of Life, a scientific detective story, (1985), Cambridge University Press

23. Kuhn H, Angew. Chemie int ed (1972) 11,798-824

24. Kuhn H, Naturwissenschaften 63, 68-80, (1976)

25. Vincent L et al in Life, 2019, vol9, p80

26. Russell MJ and Ponce A Life (2020) 10:291

27. Sojo V, Herschy B, Whicher A, Camprubi E and Lane N, The origin of Life in Alkaline Hydrothermal Vents. Astrobiology (2016) 16:181-197

28. Stuart A Kauffman. The Origins of Order, Self-organization and selection in evolution (1993), Oxford University Press

29. Shapiro Orig Life, 1984;14(1-4):565-70.

30. K Gardiner, T Marsh, N Pace, S Altman - Cell, 1983, Vol.35 (3), p.849-857

31. Forster AC et al. Nature,334, 265-267 (1998) Self-cleaving viroid and newt RNAs may only be active as dimer

32. Bowman JC, Hud NV, and Williams LD. The ribosome challenge to the RNA world. J Mol. Evol 2015, 80:143-161.

33. Gospodinov A and Kunnev D Universal codons with enrichment from GC to AU nucleotide composition revea;l a chronological assignment from early to late along with LUCA formation Life (2020)

34. Kunnev D and Gospodinov A Possible emergence of sequence specific RNA aminoacylation via peptide intermediary to intiate Darwinian evolution and code through origin of life. Life (2018)

35. Alva V, and Lupas AN . From ancestral peptides to designed proteins. Curr opin Structural biology (2018) 48:103-108.

36. Carter CW An alternative to the RNA world , Natural History 125:28-33 (2016)

37. Kalin Vetsigian, Carl Woese, and Nigel Goldenfeld PNAS July 11, 2006 103 (28) 10696-10701;

38. C R Woese, D H Dugre, W C Saxinger, and S A Dugre Proc Natl Acad Sci U S A. 1966 Apr; 55(4): 966–974.

39. Tropsha A, Kizer JS and Chaiken IM, (1992) *J Molec Recognition,* **5**:43-54

40. Orgel LE, Nature 358, 203-209 (1992)

41. Szathmary E, Proc Nat Acad Sci USA, 90, 9916-9920 (1993)

42. Ivanov OCH (1989) Origins of Life and Evolution of the Biosphere 19:187-198

43. Knutson V, J Biol Chem. 1988 Oct 5;263(28):14146-51.

44. Anfinson C (1973) *Science* **260**: 1903-1904

45. Chou PY and Fasman GD (1974) *Biochemistry* **13**:222-245

46. Kyte J and Doolittle RF (1982) *J Mol.Biol.* **157**: 105-132

47. Eisenhaber F, Persson B and Argos P, (1995) *Critical Reviews in Biochemistry and Molecular Biology,* **30**:1-94.

48. Hendry LB et al (1981) Proc Nat Acad Sci USA 75, 7440-7444

49. Zull et al TIBS July 1990, p257-261

FURTHER READING

1. The emergence of Life from chemical origins to synthetic biology (2006) Pier Luigi Luisi, Cambridge University Press
2. Life's Origin The beginning of biological evolution ed JW Schopf (2002) University of California Press
3. Vital Dust, the origin and evolution of life on Earth. Christian De Duve (1995) Basic books
4. The fifth miracle, Paul Davies (1998), Simon and Schuster.

Acknowledgements: the author wishes to thank emeritus A/Prof Victor Gostin of the geology dept of the University of Adelaide for useful discussions. The author wishes to thank Dr Jasmine Evergreen for beta reading the manuscript and making useful observations.